THE HEART MOUNTAIN DETACHMENT FAULT:
A CRITICAL REAPPRAISAL

Albert J. Warner and Thomas M. Bown

Copyright © 2025 Albert J. Warner and Thomas M. Bown.

All rights reserved. No part of this book may be used or reproduced by any means, graphic, electronic, or mechanical, including photocopying, recording, taping or by any information storage retrieval system without the written permission of the author except in the case of brief quotations embodied in critical articles and reviews.

Archway Publishing books may be ordered through booksellers or by contacting:

Archway Publishing
1663 Liberty Drive
Bloomington, IN 47403
www.archwaypublishing.com
844-669-3957

Because of the dynamic nature of the Internet, any web addresses or links contained in this book may have changed since publication and may no longer be valid. The views expressed in this work are solely those of the author and do not necessarily reflect the views of the publisher, and the publisher hereby disclaims any responsibility for them.

Any people depicted in stock imagery provided by Getty Images are models, and such images are being used for illustrative purposes only.
Certain stock imagery © Getty Images.

ISBN: 978-1-6657-7032-3 (sc)
ISBN: 978-1-6657-7033-0 (e)

Library of Congress Control Number: 2024926328

Print information available on the last page.

Archway Publishing rev. date: 02/03/2025

The Heart Mountain Detachment Fault—A Critical Reappraisal

Albert J. Warner[1]
Thomas M. Bown[2]
Mark E. Mathison[3]

[1]Senior Vice President of Geology (retired)—Lynx Resources Partners, Oklahoma City, OK 73116; Consulting Petroleum Geologist

[2]Department of Anthropology & Geography, Colorado State University, Fort Collins, CO 80523; Research Affiliate, Denver Museum of Nature & Science, Denver, CO 80205; Consulting Environmental Geologist

[3]College of Liberal Arts and Sciences, Iowa State University, Ames, IA 50011

Key Words: Heart Mountain detachment fault, detachment fault, landslide, sturzstrom, Absaroka Range, Yellowstone-Absaroka volcanism, Farallon Plate, megathrust earthquakes

Citation Suggestions: (1) **Warner, A.J., and Bown, T.M. 2024**. Shoshone/Sunlight/Abiathar Detachment Fault (SSADF); pp. 7-24 *in* Warner, A.J., and Bown, T.M. (eds.) 2024, The Heart Mountain Detachment Fault—A Critical Reappraisal. New York, Archway Publishing; (2) **Bown, T.M., Warner, A.J., and Mathison, M.E. 2024.** Heart Mountain/McCullough Peaks Sturzstrom (HMMPS); pp. 25-41 *in* Warner, A.J., and Bown, T.M. (eds.) 2024, The Heart Mountain Detachment Fault—A Critical Reappraisal. New York, Archway Publishing; (3) **Bown, T.M., and Warner, A.J. 2024.** Peripheral Phenomena; pp. 42-48 *in* Warner, A.J. and Bown, T.M. (eds.) 2024, The Heart Mountain Detachment Fault—A Critical Reappraisal. New York, Archway Publishing.

TABLE OF CONTENTS

TECHNICAL ABSTRACT 1

INTRODUCTION 3

CONCEPT REVISION 5

SHOSHONE/SUNLIGHT/ABIATHAR DETACHMENT FAULT (SSADF)
(A.J. Warner and T.M. Bown) 7

HEART MOUNTAIN/McCULLOUGH PEAKS STURZSTROM (HMMPS)
(T.M. Bown, A.J. Warner, and M.E. Mathison) 25

PERIPHERAL PHENOMENA
(T.M. Bown and A.J. Warner) 42

ACKNOWLEDGMENTS 49

REFERENCES CITED 50

TABLE I 67

TABLE II 69

TABLE III 70

TABLE IV 72

TABLE V 75

APPENDIX 76

FIGURE CAPTIONS 77

FIGURES 1-63 84

ABOUT THE AUTHORS 121

TECHNICAL ABSTRACT

The areal extent, mechanics, and timing of emplacement of the Heart Mountain Detachment Fault (HMDF) of northwest Wyoming are re-examined in light of new field evidence, incorporation of knowledge about the nature of the earth's most powerful earthquakes and largest landslides, and a reinterpretation of the geologic evidence presented in several past studies. These data indicate that phenomena attributed by all past authors to the HMDF are the result of two distinct geologic events: (1) a massive detachment fault triggered and sustained by a megathrust earthquake precipitated by late early Eocene (~49.5 Ma) plate tectonic movements that also instigated the eruptive activity and voluminous volcanic and volcaniclastic deposition of the Sunlight Group of the Absaroka Volcanic Supergroup (here termed the Shoshone/Sunlight/Abiathar Detachment Fault, or SSADF); and (2) an immense sturzstrom landslide precipitated by sustained intense earthquake activity accompanying extremely violent silicic eruptions (the Heart Mountain/McCullough Peaks Sturzstrom, or HMMPS—the world's largest known landslide) during a period of heightened regional erosion in the early Pleistocene (~2.08 Ma). Both displacements were accompanied by significant volcanic activity, with that resulting in the HMMPS being cataclysmic, and both displacements can, with a reasonable degree of certainty, be linked to specific eruptive episodes.

The sequence of SSADF events began with a brief period of late early Eocene volcanism that resulted in deposition of the Cathedral Cliffs Formation (= early acid breccia) followed by a massive bedding rupture within the basal Bighorn Dolomite and, immediately thereafter, by exceptionally violent and voluminous volcanic activity. Both episodes of volcanism were likely triggered by arc magmatism from either: 1) Farallon Plate rollback beneath the western portion of the North American Plate; or 2) a slab window between the Farallon and Resurrection Plates as flat to shallow subduction was ongoing beneath the North American Plate. Both instances of volcanism were the result of upwelling of hot aesthenospheric mantle that melted the overlying continental lithosphere of the North American Plate and/or the subducted edges of the oceanic plates. The rupture within the basal part of the Bighorn Dolomite was instantaneous and precipitated by a megathrust earthquake resulting from movement between the subducting plate(s) and the overriding North American Plate. Development of the sub-horizontal rupture may have been assisted by dynamic and thermal uplift that resulted in surface doming that accompanied magma chamber filling and produced a detached 460-610+ meter-thick upper plate carapace of largely Ordovician through Mississippian carbonate rocks and the overlying Cathedral Cliffs volcanics. Intense seismic activity of an Mw 9.0+ megathrust earthquake lasting several minutes instigated the subsequent catastrophic "breakaway" of this rupture-separated, carapace-like upper plate and its rapid (150-1,224 kmph) propulsion about 25 km to the south and southeast down a sloping surface of about 2°. This surface was overridden by a continuous though fractured allochthon that quickly broke up into individual blocks that were dispersed rapidly down slope, resulting in a vast surface of tectonic denudation. A number of the massive blocks rotated about a vertical axis during movement.

High-pressure upper plate/lower plate contact phenomena indicate a catastrophic breakaway and early rapid displacement of an intact, continuous allochthon and the probable short-term reduction of friction by chemical and/or hydrostatic means; however, the rapid dismemberment of the mobile allochthon into individual blocks forced the escape of any pressurized gases/fluids. Movement of the isolated allochthons was maintained by sustained Mw 9.0+ megathrust earthquake seismicity for 4-10+ minutes that essentially reduced to zero the coefficient of friction between the upper and lower plates. This scenario strongly supports E.H. Stevens' and W.G. Pierce's concept of catastrophic movement of individual blocks forming and transgressing a "surface of tectonic denudation" and argues against the involvement of any post-Cathedral Cliffs volcanic rocks (*i.e.*, Wapiti Formation) in detachment fault movement. The downslope advance of the disjunct Paleozoic allochthons slowed at the eastern edge of the Sunlight Basin as the allochthons piled up at the base of the upward-inclined transgressive ramp; however, several SSADF blocks ascended the ramp and accumulated on the former land surface between Pat O'Hara Mountain and Dead Indian Hill (Natural Corral area), and above the juncture of Rattlesnake and Pat O'Hara mountains.

The late early Eocene detachment fault phenomena distributed between Abiathar Peak and the Sunlight Basin/Shoshone River area are given the new name Shoshone/Sunlight/Abiathar Detachment Fault (SSADF). This detachment faulting was immediately succeeded by the rapid deposition of up to 2,000 m of volcanic rocks and their volcaniclastic equivalents (largely Wapiti Formation) that buried most of the vast area covered by allochthons and an extensive surface of tectonic denudation. Paleotopographic and other considerations contradict the recently proposed inclusion of allochthonous volcanic rocks atop Squaw Buttes in SSADF activity, as well as belie a suggestion that the SSADF deposit dammed a hypothetical, southerly flowing drainage that flowed from the Absaroka Range to southwest Wyoming and restricted deposition of volcanic sediments in middle Eocene Lake Gosiute.

Geomorphological studies demonstrate that all allochthonous Paleozoic rocks on Heart Mountain proper, on McCullough Peaks, and those distributed over intervening and outlying areas were derived from SSADF upper plate rocks that made it over the ramp onto the former land surface, where they remained poised high above the eastern edge of the Sunlight Basin (at the western margin of the Bighorn Basin) for the remainder of the Tertiary. They owe their present locations to a massive sturzstrom landslide (Heart Mountain/McCullough Peaks Sturzstrom = HMMPS) set in motion by the exceedingly violent ~2.08 Ma Huckleberry Ridge silicic eruptions and the accompanying sustained earthquakes—the most powerful volcanic activity in the geologic history of the Absaroka Range. This landslide took place during an episode of extensive regional erosion that has lasted to the present day, and one that removed most of the landslide deposit. HMMPS landslide debris formed natural dams that diverted the north-flowing Shoshone River to the south and east through the Oregon Basin (a dissected structural dome) where it rejuvenated the drainage of Dry Creek, a stream course earlier abandoned through its capture by the Greybull River. Subsequent headward erosion penetrated the landslide dams, re-established the northerly course of the Shoshone River, and resulted in a second abandonment of the channel of Dry Creek.

INTRODUCTION

Previous Work

The Heart Mountain Detachment Fault (HMDF) of the northwestern Bighorn Basin, Wyoming (**Figure 1**), is an enigmatic structure, several aspects of which have puzzled geologists for more than one and a quarter century (Eldridge, 1894; Fisher, 1906). As regards the detachment and movement of the upper plate, Davis (1965) opined that it is "...difficult to comprehend by any known mechanism," a conclusion reached by nearly every student of the displacement. Previous research on the HMDF may be broadly categorized in five somewhat temporally overlapping stages: (1) recognition and definition (studies of Eldridge, Fisher, Dake, Hewett, Stevens, and Bucher, 1894-1947); (2) W.G. Pierce's and colleagues' detailed geologic mapping of the northeastern Absaroka Range and Pierce's development of the tectonic denudation model (1941-1993); (3) T.A. Hauge and the continuous allochthon model (1983-2011); (4) numerous authors' detection of various autochthon/allochthon contact phenomena resulting from physico-chemical changes that likely reduced friction beneath an initially intact, mobile HMDF allochthon to a greater or lesser degree for greater or lesser periods of time (1933-2018); and (5) D.H. Malone and his colleagues' refinements of dating of HMDF phenomena, and their proposed extensions of: a) the areal distribution of HMDF (= SSADF) allochthons 50 miles (82 km) southeastward to Squaw Buttes; and b) the effects of HMDF (= SSADF) faulting to sedimentation in Lake Gosiute in southwest Wyoming (2007-2014).

Stages 1 and 2 of HMDF study consist principally of the compilation of geologic mapping and other information derived from field observations that effectively constrained interpretations of the nature of the faulting such that, by about 1980, aside from the development of a reliable and detailed geochronological dating of the sequence of detachment faulting events, the principal remaining problem of the HMDF seemed to be the nature of the triggering mechanism and how fault movement was sustained, including the enigma of by what means friction along the fault plane within the Bighorn Dolomite was reduced such that the titanic allochthons of Paleozoic rocks (many of them several km^3 in volume) were transported tens of kilometers across an autochthonous lower plate rock surface sloping 2° or less to the southeast.

The studies of W.G. Pierce and his co-authors provide insightful, well-reasoned explanations for nearly all Heart Mountain detachment fault phenomena and, following our seven seasons of fieldwork, we conclude that Hauge's continuous allochthon extensional gravity-spreading hypothesis, Malone and his co-authors' expansion of Heart Mountain faulting phenomena to Squaw Buttes, and the attribution of the effects of that faulting to aspects of lacustrine sedimentation in southwest Wyoming are based on very little—and of that, arguable—field evidence. Their interpretations have generated a great many additional questions and provided few, if any, solutions to old problems.

In proposing that massive volumes of volcanic rocks (largely Wapiti Formation) were involved in HMDF (= SSADF) faulting, Hauge's model of extensional gravity spreading of a continuous allochthon not only obviates Pierce's model of tectonic denudation but also introduces a plethora of new questions that seem to defy resolution and that will be discussed in detail below. Although we applaud Malone and his co-workers' refinements of the dating of certain HMDF and related phenomena (*e.g.,* Malone, Craddock, and Mathesin, 2014; Malone, Craddock *et al.*, 2015), the geologic evidence does not argue convincingly for his HMDF origin of allochthonous volcaniclastic rocks capping Squaw Buttes, nor does it support the idea that damming by HMDF allochthons of a hypothetical, southerly flowing Eocene Bighorn Basin drainage affected sedimentation in middle Eocene Lake Gosiute in southwest Wyoming.

Abbreviations: HMDF = Heart Mountain Detachment Fault (all early Eocene detachment fault phenomena given that name and as conceived by all previous authors); SSADF = Shoshone/Sunlight/Abiathar Detachment Fault (our construct of the late early Eocene part of the HMDF, minus the allochthons on Heart Mountain, on McCullough Peaks, and in intervening areas); HMMPS = Heart Mountain/McCullough Peaks Sturzstrom (early Pleistocene landslide deposits earlier identified as the easternmost remnants of the HMDF and including all allochthonous masses of Paleozoic carbonate rocks on Heart Mountain, on McCullough Peaks, and in several intervening and outlying areas); ECOCDA = Enos Creek/Owl Creek Debris-avalanche of Bown (1982a,b) and Bown and Love (1987).

CONCEPT REVISION

Omitting the terms "rock glacier," and "circular fault" which clearly do not apply, and the nondescript appellation "allochthon," the following genetic designations have been applied to the HMDF by various authors (listed in no particular order): a detachment fault; a landslide; a slide block; a detachment; a block slide; the Heart Mountain Thrust; the Heart Mountain allochthon; Heart Mountain faulting; the Hart [sic] Mountain Overthrust; the Heart Mountain slide; a detachment thrust; and the Heart Mountain rockslide. These constructs may be reduced to two distinct concepts: (1) a detachment fault, a detachment, a thrust, faulting, an overthrust, a detachment thrust; and (2) a landslide, a slide block, a block slide, a rockslide, and a slide. Further reduction gives us: (1) a detachment fault, and (2) a landslide.

Thorough reexamination of the field evidence shows that the structural feature most commonly termed the Heart Mountain Detachment Fault (HMDF) is actually the conflation of two geologic structures of distinctly different ages and origins. The older we term the Shoshone/Sunlight/Abiathar Detachment Fault (new name=SSADF) of late early Eocene age; and to the younger we apply the name Heart Mountain/McCullough Peaks Sturzstrom (new name=HMMPS) of early Pleistocene age (an immense landslide).

Pierce (1973:465) clarified the distinction between a detachment fault and a landslide deposit, observing that:

"… material making up the landslides is not mapped as transported formations but has become a new mappable unit."

Rocks of the SSADF are clearly distinguished as transported formations, whereas those of the HMMPS deposit comprise a new, mappable unit made up of the jumbled remains of several Paleozoic units. Davies *et al.* (1999:1096) supported Pierce's view regarding the makeup of landslide debris, noting that:

"… deposits of large rock avalanches [landslides] consist predominantly of intensely fragmented rock debris …" and "… fragmentation appears to occur throughout the runout."

Allochthonous rocks of the SSADF and the HMMPS deposit have been conflated as being of a single origin (HMDF) for more than a century because: (1) both structures exhibit masses of older rocks lying atop younger rocks; (2) both structures involve displacement of the Ordovician Bighorn Dolomite, the Devonian Jefferson and Three Forks formations, and the Mississippian Madison Limestone; and (3) both structures are developed in the same regional area and are seemingly contiguous, such that the HMMPS deposit appears to be the easternmost extension of the SSADF. Even the degree of seemingly progressive deformation of allochthonous rocks from northwest to southeast—from the castellate escarpment of relatively undeformed Paleozoic strata at Cathedral Cliffs (nearer the "breakaway" site—**Figure 2**) to the jumbled, tilted blocks on and at the base of Pierce's "ramp" (**Figure 3**) to the largely comminuted carbonate debris atop McCullough Peaks (**Figure 4**)—appears to corroborate a single displacement origin.

Poor outcrop patterns, differential preservation of allochthonous debris, and differing access to the many important outcrops in the vast study area by interested scientists, American and foreign, has also contributed to confusion. Whereas allochthons of the SSADF were transported by plate tectonic-induced and sustained megathrust earthquake seismicity during a massive and violent volcanic *depositional regime* (Sunlight Group of Absaroka Volcanic Supergroup) and have been largely buried by nearly 2,000 meters of volcanic rocks (Rouse, 1937), the landslide debris of the early Pleistocene HMMPS (triggered by cataclysmic silicic volcanism accompanied by sustained earthquake activity) was transported during a major *erosional regime* and the preponderant volume of that debris was removed by stream erosion subsequent to landsliding.

SHOSHONE/SUNLIGHT/ABIATHAR DETACHMENT FAULT (SSADF)
(A.J. Warner and T.M. Bown)

Stratigraphic Setting

The greater Absaroka Volcanic Field was active from about 53-43 Ma (Hiza, 1999a); *i.e.*, from the early Eocene through the middle part of the middle Eocene. The Absaroka Volcanic Supergroup (Smedes and Prostka, 1972; Fig. 3) is a body of volcanic and volcaniclastic rock up to 2 km thick (Rouse, 1937:1262), about 180 miles (295 km) long and 60 miles (~100 km) wide (29,500 km^2), extending NW-SE from just west of Livingston, MT, to just south and east of Dubois, WY. The Supergroup is subdivided into three groups developed in two subparallel trends which, oldest to youngest and from NW to SE, are the Washburn, Sunlight, and Thorofare Creek Groups (Smedes and Prostka, 1972). The Washburn Group is located primarily in Montana, west and southwest of Livingston, and extends into northeastern Yellowstone National Park. The Sunlight Group outcrops from the Beartooth Range in Montana southeast to just west of the Buffalo Bill Reservoir (west of Cody, WY), and then south to just east of Dubois, WY. Much of the southeast portion of the Sunlight Group is covered by the younger Thorofare Creek Group, which occupies the area from northeastern Yellowstone National Park to Dubois. The Sunlight Group is exposed in a narrow band along the eastern margin of the Thorofare Creek Group southwest of Buffalo Bill Reservoir.

This study is largely concerned with catastrophic geological events that occurred: 1) immediately preceding deposition of the Sunlight Group (late early Eocene); and 2) in the early Pleistocene.

William G. Pierce Interpretation

Hypotheses concerning the origin and nature of emplacement of allochthons attributed to the Heart Mountain Detachment Fault (HMDF) have coalesced today into two competing concepts. The first, advocated by William G. Pierce for more than three-quarters of a century, built upon earlier work by Bucher (1933, 1947), who envisioned a catastrophic, earthquake oscillation origin for the faulting, and Stevens (1938) who was the first to suggest the upper plate moved as several individual blocks of a fractured allochthon. Through a prodigious, decades-long program of geologic mapping (Pierce, 1965a, 1965b, 1966a, 1978, 1985, 1997; Pierce and Nelson, 1969, 1970; and Pierce, Nelson, and Prostka, 1973, 1982), Pierce was the first to delineate the magnitude of the detachment faulting and to identify: 1) the breakaway site of the detachment (**Figure 5**) (Pierce, 1960, 1980); 2) a surface of tectonic denudation (**Figure 2**) produced by dismemberment and dislocation/evacuation of the upper plate (Pierce, 1957, 1968a, 1987; Pierce and Nelson, 1986); 3) a transgressive fault (Pierce, 1941, 1957, 1960) and ramp upon which one massive and several smaller allochthonous bodies are preserved (**Figure 3A**); and 4) a former (Eocene) land surface (Pierce, 1960, our **Figure 6**) traversed by some of the detached masses.

Pierce argued that field evidence demonstrated the emplacement of the detached masses took place "near the close of the early Eocene" (Pierce, 1973:462), prior to deposition of the Wapiti Formation (Pierce, 1973, 1982; Prostka, 1978), and that movement was both catastrophic and rapid. With some elaboration and a few modifications, we are in general agreement with Pierce's interpretation of the origin, nature, and speed of early Eocene detachment faulting that he termed the Heart Mountain Detachment Fault (our SSADF). Pierce (1973) published an excellent summary of his views on this hitherto enigmatic structure, and Pierce's (1973) cogent responses to other geologists' comments on the "mechanism problem" remain in force today.

Thomas A. Hauge Interpretation

Hares (1934) first proposed that Paleozoic allochthons of the HMDF (our SSADF) were held together during movement by volcanic agglomerate, a concept resurrected in some measure by Hauge (1982, 1983, 1990b) and more recently espoused by Malone (1995) and Beutner and Craven (1996). Malone et al. (2014:339) asserted that Pierce's (1973) tectonic denudation model of Heart Mountain Detachment Fault movement "... has been largely abandoned as a viable model" in favor of Hauge's (1982, 1983, 1985, 1990b, 1993) extensional, gravity-spreading (= continuous allochthon) model, in which an immense volume of volcanic rocks (including all or part of the Wapiti Formation) is recognized as the overwhelmingly principal component of the upper plate (see also Beutner and Hauge, 2009). Hauge (1982) also rejected the concept of a rapid, catastrophic emplacement of the HMDF allochthons, ultimately considering that their emplacement took place "... over a million or more years..." (see also Templeton et al., 1995). Hiza (1999b) later suggested that HMDF emplacement required >2 Ma. Beutner and Hauge (2009) concluded that not only were Wapiti rocks incorporated in the HMDF allochthon, but that some movement probably took place after what they perceived to have been the collapse of an "Eocene volcanic system" (see, e.g., Francis and Self, 2020).

In contrast, Malone and Craddock (2008) maintained that allochthon ("upper plate") movement was catastrophic, and Malone et al. (2014) went on to assert that:

> "The consensus among those working on the slide today is that it was a catastrophic event associated with edifice collapse in the Eocene Absaroka volcanic province ..."

Although it is perhaps easier to visualize the "collapse" of a volcanic system/edifice to be catastrophic in nature rather than attenuated over 1-2 million years, Hauge and Hiza see the movement as slow, whereas Malone and most other students of the HMDF phenomenon have argued for a catastrophic origin and rapid movement of the displaced masses.

In his construct of the detachment, Hauge did not clearly define a "breakaway" phenomenon associated with any particular volcanic activity, so we are left assuming that HMDF movement began at some point during or near the end of Wapiti deposition, for otherwise, in the pertinent area of the Absaroka Range, there would have been no volcanic edifice to collapse. Malone and Craddock (2008:21) attribute fault movement to "hydrothermal systems" injecting "volcanic gas and glass" and the "heating of pore waters through volcanic intrusion." Certainly, such specific attributes, for which there is absolutely no field evidence, would more likely have obtained during volcanism that took place during Wapiti deposition rather than afterward.

Hauge (1985, 1990a, 1990b) and Beutner and Gerbi (2005) believe the Paleozoic part of the allochthon broke apart after it was entombed by volcanics. This is an extraordinary concept as it would require that, in several instances (Malone et al., 2017; Malone et al., 2022), the independent rotation of massive blocks of Paleozoic rocks up to 35° occurred *after they were engulfed by volcanic rocks*. This and other signal problems with Hauge's interpretation are discussed further below.

Shoshone/Sunlight/Abiathar Detachment Fault: The Upper Plate
Composition

By our interpretation, the majority volume of the upper plate of the SSADF consists of 460-610+ meters of Lower Paleozoic rocks, largely carbonates, of the Ordovician Bighorn Formation, the Devonian Jefferson and Three Forks Formations, and the Mississippian Madison Formation (Pierce, 1957, 1963b, 1973). According to Pierce and Nelson, 1973:2631), at some time in the early Eocene, and prior to any volcanic deposition "... preliminary movement of the Heart Mountain detachment fault ..." opened a single, deep, narrow rift in the upper plate Paleozoic rocks along which a stream cut a channel deep into the underlying Cambrian Snowy Range Formation. The channel filled with conglomerate (= Crandall Conglomerate—Pierce and Nelson, 1973) formed of clasts derived almost exclusively from Paleozoic carbonates making up the bounding canyon walls. The conglomerate was then lithified and its upper part—that deposited at a level above the base of the Bighorn Dolomite—was truncated and transported as part of the upper plate of the Heart Mountain detachment fault (=SSADF). Following deposition of the Crandall Conglomerate but prior to SSADF movement, 150-275 meters of the volcanic and volcaniclastic Cathedral Cliffs Formation (= early acid breccia of Hague, 1899) was deposited (Pierce, 1963a), and rocks of both the Madison Formation and Cathedral Cliffs Formation were displaced by the Reef Creek detachment fault (Pierce, 1963b).

We agree with Hauge (1990) and Beutner and DiBenedetto (2003) that the Crandall Conglomerate was probably deposited in one or more non-tectonic, stream-eroded canyons prior to any SSADF movement. However, in contrast with Hauge, we contend that the only volcanic and volcaniclastic rocks transported by Heart Mountain (SSADF) detachment faulting are some rocks of the Cathedral Cliffs Formation, including those displaced earlier by the Reef Creek Detachment Fault (Pierce, 1963b), and those rocks rode "piggy-back" atop SSADF allochthons (*e.g.*, Pierce, 1963a, Fig. 2).

Discussion

Hauge (1982, 1983, 1985, 1990a, 1990b), Malone (1995), and Beutner and Craven (1996), among others, believed huge volumes of volcanic rocks (including most or all of the Wapiti Formation) to be a major constituent of the upper plate. The Hauge model of extensional gravity-spreading of a titanic, 1-2 km-thick, sheet-like allochthon composed largely of volcanic and volcaniclastic rocks, obfuscates without resolution much complicated geology thought to have been settled by Pierce's detailed mapping and interpretations. Even Malone's (2000) more restricted interpretation, in which he regards his (1995) Deer Creek Member of the basal Wapiti Formation to be the only detached volcanic unit, introduces a plethora of new questions. Chief among the unaddressed/unexplained problems with Hauge's continuous allochthon model are:

1) How far to the east did the Absaroka volcanics extend prior to detachment faulting (see discussion by Ritter, 1975), and where was the toe of Hauge's continuous allochthon detachment fault? Was the toe at the edge of the volcanic field, or was there a transgressive fault through the volcanics? What thickness of Wapiti Formation was displaced? Hauge (1985) suggested that the thickness of displaced Wapiti rocks was up to 2 km, and later (1993, Figure 1), that the thickness was at least 1 km.

2) How does the scenario of gravity-induced tensional normal fault extension of a continuous allochthon explain Paleozoic allochthons that climbed ~1,300 feet (~397 m) up the ramp between Pierce's area of tectonic denudation and the ancient Eocene land surface; *i.e.*, how was *gravitational* extension accomplished *uphill*? Also, as observed by Pierce (1973:463):

> "... along the transgressive fault, upper plate blocks rest on the Heart Mountain fault where it slopes as much as 10°, but they are not now sliding under the influence of gravity."

3) Hauge identifies very few major normal faults in his scenario of a continuous volcanic allochthon, yet they play a seminal role in his model. *Where are all the faults?* There should be a plethora of them and, because most of them (according to his model) must run for tens of miles and displace a minimum of 1,000 m of volcanic rock, at least one of them should be obvious. Such faults are structural features that could be documented by mapping, yet Hauge offers no new mapping and describes only a few, relatively minor faults. Neither we nor Pierce (1973) have been able to visualize those in the field as faults as opposed to depositional contacts of volcanics against Paleozoic carbonate rocks. For extensive listric normal faulting to have been the mechanism by means of which a kilometer-thick (or more) package of Wapiti Formation and 460-610+ meters of Paleozoic carbonates were transported tens of miles over 1-2 million years, dozens—if not hundreds—of such faults should be *clearly* expressed on a titanic scale, mappable over wide areas, and at least some fault contacts of volcanics with Paleozoic rocks should be explicit. Where are the slickensides? Where is the evidence of displacement? *Where are the faults?*

Hauge (1993:542) stated:

> "Commonly ... stratification is difficult to discern in volcanic rocks above the detachment ... This lack of readily apparent stratification of Eocene strata has greatly hindered understanding of the extent of involvement of volcanic rocks in Heart Mountain faulting and the amount of internal extension of these rocks."

We suggest these enormous faults are difficult to discern because they do not exist. For example, a section of 2,000 - 3,300 feet (610 - 1,000 m) of well-stratified Wapiti Formation rocks is superbly exposed for several miles east and west of Jim Mountain (peak at 44° 31' 43.42" N, 109° 28' 27.67" W), in the Wapiti Valley west of Cody, WY (*e.g.*, **Figures 7 and 8**). At no Place along this titanic, castellate body of rock is exposed even a single fault of remotely near the magnitude necessary to have facilitated displacement by the continuous allochthon hypothesis. Nor are they to be seen in more remote areas, all of which are now accessible through remote sensing. Furthermore, the numerous, huge listric faults Hauge proposed should exhibit at least some backward rotation of the downfaulted blocks, yet no such relationships are seen.

4) Where is the volcanic breccia beneath (overridden by) the Paleozoic allochthons? Given Hauge's view that the overwhelming volume of the allochthon is volcanic, such breccia should not only be present, it should be ubiquitous. Hauge's pre-detachment volcanics (including all or part of the Wapiti Formation) must have blanketed an immense area east of the Paleozoic masses and, at the onset of movement, the toe of the detachment must have been topographically much lower (down the 1-2° dip surface) than the widespread, planar fault surface in the Bighorn Dolomite, and did not involve Paleozoic rocks. If the toe of the volcanic part of the continuous allochthon lay at or near the base of the volcanic rocks, how and

why did the detachment fault project upwards and to the northwest to form a widespread, planiform fault surface at a horizon near the base of the Bighorn Dolomite? If that is what transpired, once movement was initiated, the Paleozoic parts of the allochthon would have overridden an immense area of truncated volcanic rocks, yet no such field relationships exist.

Gerbi (1997:146) mistakenly stated that: "A breccia layer exists along much of the fault …". Like many geologists studying the contact relationships of the HMDF, Gerbi based his conclusions chiefly on data collected at White Mountain, a locality that, if volcanic rocks are presumed to have made up 70-90% of the volume of the upper plate (Hauge, 1985), exhibits some of the least representative contact phenomena.

The style and intensity of deformation of what actually are catastrophically displaced volcaniclastic rocks in the Absaroka Range (**Figure 9**) is evident in the transported debris of the Enos Creek-Owl Creek debris-avalanche (=ECOCDA; Wilson, 1975a; Bown, 1982a, 1982b; Bown and Love, 1987) and the Deer Creek debris-avalanche (Malone, 1995, 2000). If huge volumes of volcaniclastic rocks were transported by the HMDF (SSADF), and especially if they were transported rapidly and catastrophically, as upper plate/lower plate SSADF contact phenomena indicate, why is there so little deformation at the base and, at least, in the lower part of the supposedly allochthonous volcaniclastics (*i.e.* the Wapiti Formation)? Such deformation of volcanics as does exist regionally is readily explained by: 1) movement associated with the Deer Creek debris-avalanche (affecting rocks in the lower part of the Wapiti Formation according to Malone, 1995—but actually deforming pre-Wapiti and pre-SSADF affected Cathedral Cliffs Formation and equivalents); 2) by the "chaotic dismemberment" described by Decker (1990—our **Figure 10**—see also chaotic tuff-breccia of Hay, 1954); and/or 3) mantle bedding (*e.g.*, Bown and Larriestra, 1990). We agree with Decker that liquefaction-related deformation of Cathedral Cliffs and Aycross equivalent (pre-Wapiti) strata in the valleys of the North and South Forks of the Shoshone River (**Figure 11**) and the valley of the Greybull River (*e.g.*, southwest of Rose Butte, **Figure 12**) was probably coincident with HMDF (=SSADF) movement. It remains to be determined just what is the difference between deposits of Malone's Deer Creek debris-avalanche and rocks affected by Decker's chaotic dismemberment.

Hauge's (1985: Figures 4 and 5) illustrations of Hauge's (1993:542): " … deformed base of volcanic rocks …" that are alleged to show a similar range of thickness as Pierce's (1979) 15 m or more of deformed Paleozoic rocks above the Paleozoic autochthon are close-up images lacking scale that illustrate only slightly deformed volcanic rocks of indeterminate stratigraphic position that might equally well owe their "deformation" (poorly expressed in Hauge, 1985, Figure 4; simple dipping strata in Hauge, 1985, Figure 5) to movement associated with Malone's (1995) Deer Creek debris-avalanche. They are wholly inadequate as documentation of gouge of volcanic/volcaniclastic rocks or as deformation at or near the base of a massive, 1,000+ meter-thick volcanic/volcaniclastic allochthon occupying an appreciably significant volume of the northeast Absaroka Range.

5) What post-Wapiti seismic/volcanic event triggered the allochthonous movement of Wapiti rocks? If movement was sustained by simple gravity pulling an enormous volume of rock down a slope of only 1-2° for 1-2 million years, why aren't phenomena like the SSADF more common, especially as thick accumulations of volcanics are commonly deposited on much steeper slopes than 1-2° in the Absaroka Range and in many other volcanic fields?

6) Where is the "breakaway" scarp for the volcanic allochthon? Is one needed? Where is the succession of west-dipping, backward-rotated blocks of Wapiti Formation demanded by a mobile mass broken up and transported eastward by numerous huge listric faults?

7) If buried beneath hundreds of meters of Wapiti volcanics prior to movement, how did the Paleozoic part of the allochthon at White Mountain become independently rotated 35° (Malone *et al.*, 2017)? The same problem pertains to several other rotated allochthonous Paleozoic blocks on the Dead Indian Hill area ramp and on the former land surface (*e.g.*, Malone *et al.*, 2022).

8) By Hauge's continuous allochthon hypothesis, breakup of the Paleozoic part of the allochthon was accomplished by down-dip movement of Wapiti volcanic rocks along listric normal faults over a million years or more, and it was movement along those faults that drew the Paleozoic blocks apart. If this scenario is correct, *by what mechanism were some of the Paleozoic blocks pushed together again* and bunched up without any intervening volcanic rocks, and how did blocks exhibiting differing structural attitudes become imbricated at the southeastern margin of Sheep Mountain (**Figure 13**) and the eastern part of White Mountain (**Figure 14**)? Also, listric faulting would produce a succession of massive, backward-rotated blocks of the Wapiti Formation hundreds of meters thick. Where are they? Not one has been described.

9) Malone and Craddock (2008:21) stated that: "The consensus of work in the last 15 years indicates that volcanic rocks overlying the HMDF are everywhere allochthonous." A *consensus*, if one actually exists, *indicates nothing* whatsoever. In contrast with Hauge, Malone and his colleagues maintain that movement was catastrophic and rapid, exceeding 150 meters/second (= ~540 km/hr = ~335 miles/hr; Malone and Craddock, 2008), or 100 meters/second (= ~362 km/hr = ~225 miles/hr; Craddock *et al.*, 2009; Malone *et al.*, 2017). It is difficult to imagine a body of Wapiti volcanics several hundred square miles in area and one or two kilometers thick hurtling down a slope of 1-2° at such velocities, and one producing virtually no contact deformation.

10) Hauge's (1990) Paleozoic allochthons #2 and #21 are dipping up to 23°. If these and other inclined Paleozoic allochthons were entombed in volcanics, and their contacts with those volcanics are normal faults, what caused the dips when the underlying Pilgrim Formation dips only 1-2°? Why are none of the massive, confining volcanic parts of the allochthon in those areas (*i.e.*, Wapiti Formation) similarly inclined?

11) The area of the original, reconstructed pre-detachment mass of Paleozoic rocks is considerably less than the area over which the allochthonous Paleozoic blocks are distributed. Hauge attributes this distribution not to movement but to pre-detachment erosion, even though he (1985, 1990a, 1990b) claimed the Paleozoic part of the allochthon broke apart and moved after it was covered by volcanic rocks. Why did this erosion not somewhere penetrate the level of Pierce's "surface of tectonic denudation"? By Hauge's construct, the volume and distribution of allochthonous volcanic rocks vastly exceeds that of the Paleozoic masses, yet there exists no evidence of either fault contact brecciation of volcanic rocks beneath the allochthon or upper plate/lower plate solution/fluid pressure contact phenomena at the base of the overwhelmingly greater volume of volcanic rocks such as those described by many authors for the contact between upper/lower plate Paleozoic rocks.

In sum, we contend that: 1) the juxtaposition of several immense, variously dipping Paleozoic allochthonous blocks at the boundary between what Pierce has identified as the surface of tectonic denudation and his ramp onto the ancient land surface at the eastern edge of the Sunlight Basin (Pierce and Nelson's 1973:2637 "... scattered blocks ... bunched together as other blocks behind them continued moving down the bedding fault"—Pierce, 1957, Figure 4); and 2), the shingled imbrication of Paleozoic allochthons on the south side of Sheep Mountain (**Figures 13** and **15**) and at White Mountain (**Figure 14**), like splays of playing cards, and similar relationships in other areas argue forcefully in support of Stevens' and Pierce's hypothesis of movement of individual Paleozoic blocks across a surface of tectonic denudation prior to the blocks being covered with volcanic rocks. Otherwise, how was the rotation of massive blocks of Paleozoic carbonate rocks achieved if those blocks were entombed in flat-lying Eocene volcanics?

Several other lines of evidence support the tectonic denudation model: 1) Pierce (1963:1229) observed that:

> "... *had the early basic breccia* [= Wapiti Formation] *been involved in the faulting, one would expect to find a few fragments of volcanic rock along the fault beneath some of the limestone fault masses but none have been found.*" Brackets are ours.

Carbonate fault breccia containing no volcanic rock is additional evidence that the contact of the fault surface with the overlying Wapiti Formation is depositional (Pierce *et al.*, 1991). 2) As noted by Pierce (1973, 1987), the lack of erosion on the exposed fault plane (= surface of tectonic denudation) is compelling evidence that little time elapsed between SSADF movement and the onset of deposition of volcanic rocks of the Wapiti Formation, and argues forcefully for his tectonic denudation model. 3) The breakaway fault (Pierce, 1960, 1980) truncates the Cathedral Cliffs Formation and is abutted by and overlain by the younger Wapiti Formation (**Figure 5**) —a unit that is not cut by the SSADF. This relationship demonstrates clearly that the SSADF pre-dates Wapiti deposition (Pierce, 1982). Conveniently, Hauge (1985, 1993), in contrast with all other HMDF studies, side-steps this problem by claiming, quixotically, that rocks of the Cathedral Cliffs and Wapiti Formations cannot be differentiated (see discussion below). 4) Craddock *et al.* (2009) report horizontal rotation about a vertical axis of 113° (clockwise) or 247° (counterclockwise) of the block of marbleized Madison Limestone at White Mountain. How, according to Hauge's continuous allochthon hypothesis, was rotation of the block accomplished if it was encased in volcanic rock? Malone *et al.* (2022) described the rotation of other blocks of Paleozoic carbonate rocks but failed to discuss the significance of this relationship.

The only exposed part of Pierce's (1960) breakaway scarp (**Figure 5**) lies in the extreme northwest part of the HMDF (SSADF) faulted area, just west of Silvergate, MT, leaving open the possibility that the pre-breakaway exposure of upper plate Lower Paleozoic rocks might have been a peninsula-like erosional salient in map view; *i.e.*, the "toe" of the detachment was erosional rather than tectonic, and the margins of the allochthon overrode an eroded land surface rather than developed as a transgressive, thrust-like fault, in which Bighorn Dolomite overrode Bighorn Dolomite, as envisioned by Pierce (1957, Fig. 16; 1960, Fig. 106.1). This view is consistent with: 1) the presence of upper plate Crandall Conglomerate occurring to the west of autochthonous exposures, and 2) the most distal Logan Mountain/Sheep Mountain allochthon(s) being developed east-southeast of Sunlight Volcano.

Conversely, there need not have been a fault "toe" of the upper plate at all; the concept of a "toe" and unidirectional movement being a holdover from analogy of the HMDF structure with a thrust fault. Rather, it seems more likely that as the detached carapace broke apart from front-to-back its isolated component blocks scattered radially in all downslope directions—many, if not all of the blocks rotating somewhat as they skittered jerkily across the tectonically denuded autochthon and up the ramp. Hauge's (1993) view that the "toe" of his continuous allochthon was developed in rocks of the lower Eocene Willwood Formation is contradicted by landslide debris of the HMMPS deposit lying in clear depositional contact with Willwood rocks with no underlying "thrusting" within, or detachment of, Willwood rocks.

Triggering Mechanism and Age of Displacement

Most authors agree that movement on the Heart Mountain Detachment Fault (=SSADF) was catastrophic; *e.g.*, as proposed by Bucher (1947), and supported by Pierce (1957, 1979), and Hughes (1970) and, at least initially, the result of seismic activity resulting in explosive volcanism: *e.g.*, Melosh (1983); Beutner and Craven (1996); Craddock *et al.* (2000, 2009); Beutner and Gerbi (2005); Aharnov and Anders (2006); Beutner and Hauge (2009); Goren *et al.* (2010); Anders *et al.* (2010); Malone *et al.* (2014); Losh (2018). A wide-ranging number of triggering mechanisms have been proposed that were the result of explosive volcanism (Bucher, 1933; Beutner, 2002; Beutner and Craven, 1996; Melosh, 1989, 2018; Malone *et al.*, 2017). However, and even among recent workers, scant attention has been paid to the role plate tectonics played in the instigation of this immense displacement. We suggest that a plate tectonically induced megathrust earthquake of Mw 9.0+ at about 49.5 Ma and lasting several minutes instigated and sustained movement on the SSADF. This powerful earthquake was preceded by and followed immediately by significant volcanism.

We agree with Pierce (1987) that recognition of two Eocene rock units, the Cathedral Cliffs Formation and the Wapiti Formation, is key to understanding the sequence of events that occurred before, during, and after the emplacement of the allochthon (our SSADF). The strong contrast between the two units on the basis of both appearance and composition as well as the unconformable relationship between them recognized by all earlier workers was dismissed by Hauge (1983) in his abandonment of the subdivisions of the Absaroka Volcanic Supergroup. This dismissal enabled Hauge to defend his continuous allochthon model for the HMDF (SSADF) that has been accepted subsequently by recent workers (*e.g.*, Malone and Craddock, 2008). Because the distinction between the Wapiti and Cathedral Cliffs formations is critical to our argument in support of Pierce's view of the composition of the Absaroka Volcanic Supergroup and the nature of the SSADF, a brief summary of pertinent previous work is presented here.

Hague (1899:3) defined two units that have become critical in understanding both the stratigraphy of the northeast Absaroka Range and the nature of the HMDF (SSADF); these are the "early acid breccia" and the overlying "early basic breccia". Hague described the "early acid breccia" (= the largely clastic, alluvial Aycross Formation south of Carter Mountain) as being composed of:

> *"... fragmental material consisting of agglomerates, silts, muds, and tuffs. In color they present usually light tints, varying from grayish white to lavender. Occasionally, they are greenish brown ..."*

By comparison, the "early basic breccia" (= Wapiti Formation of Nelson and Pierce, 1968) is:

> *"... in strong contrast to the early acid breccias, they are usually dark colored owing to the amount of ferromagnesian minerals in the rocks."*

Pierce (1963a) introduced the name Cathedral Cliffs Formation to replace the term "early acid breccia". In the Clark's Fork area, as viewed from the Chief Joseph Highway (Wyoming State Highway 296), Pierce (1963a) described the Cathedral Cliffs Formation as composed of tuffs with lesser amounts of volcanic sedimentary rocks and breccias that are unconformably overlain by the "early basic breccia". This unconformity is marked by blocks of Madison Limestone (**Figure 2A**) that were entrained as part of the Reef Creek Detachment Fault (of Pierce, 1963b), a displacement that took place before both the main HMDF and deposition of the "early basic breccia". The contrast in appearance is striking at this location as the Cathedral Cliffs has an overall olive to greenish-gray color, and the overlying "early basic breccia" (later defined as the Wapiti Formation by Nelson and Pierce, 1968) is dark reddish brown. Furthermore, the "early basic breccia" has a rough, ropy surface (Hague, 1899) as compared to the smoother aspect of the "early acid breccia" (=Cathedral Cliffs Formation).

Smedes and Prostka (1972) revised the rock nomenclature of the Absaroka Volcanic Field in and near Yellowstone National Park in their construct of the Absaroka Volcanic Supergroup, defining, in ascending order: the Washburn Group, Sunlight Group, and Thorofare Creek Group. The Washburn Group includes the old "early acid breccia" and consists of the Cathedral Cliffs and Lamar River formations. The Sunlight Group is equivalent to the old "early basic breccia", and includes the Wapiti Formation and the Trout Peak Trachyandesite. Smedes and Prostka observed that rocks of the Sunlight Group are the most mafic and have the highest proportion of potassic basaltic rocks, whereas rocks of the Washburn and Thorofare Creek groups are generally lighter in color and of andesitic or dacitic composition.

Inspection of the stratigraphic columns on geologic quadrangle maps covering the research area (*e.g.*, Pierce, 1965b; Pierce and Nelson, 1968, 1969, 1971; Pierce, Nelson, and Prostka, 1973, 1982) reveals that, in all instances, the lithologic distinctions between the Cathedral Cliffs Formation (= "early acid breccia") and the definitively overlying Wapiti Formation (= "early basic breccia") are not only striking but pronouncedly mappable. The Cathedral Cliffs Formation is light in color, most often a light to medium olive to greenish-gray and, less often, light brown to gray, buff, and pink. In contrast, the Wapiti Formation is reddish brown to dark brown and the unit is commonly massive and crudely to irregularly bedded with "wedge-like" masses. The Cathedral Cliffs Formation is largely composed of tuff, volcanic sandstones, siltstones, conglomerates, and breccias, with igneous rock types including andesite, latite, and rhyodacite. Other constituents include the minerals quartz, potassic feldspar, biotite, plagioclase, hornblende, augite, and opaque minerals. Primary constituents of the Wapiti Formation, on the other hand, include basaltic lava flows and massive volcanic conglomerates, lahars, and breccias dominated by andesite and basalt. Of particular interest is the presence of pyroxene and olivine—two of the most mafic minerals—in descriptions of Wapiti rocks in the above quadrangles. Neither pyroxene nor olivine, nor any basalts or basaltic lava flows, are recognized in any sections of the Cathedral Cliffs Formation.

Based on the above observations and mapping studies by several geologists and our own field studies, we surmise that Hauge ignored nearly 90 years of detailed stratigraphic and mapping studies to lump rocks of the Cathedral Cliffs and Wapiti formations together as an inseparable unit in support of his continuous allochthon model for movement on the HMDF. See also Nelson (1991), who concluded that:

> "... *recognition that post-faulting volcanic rocks can be distinguished from pre-faulting volcanic rocks provided additional support for the tectonic denudation model.*"

Bucher (1933) first suggested that a large volcanic explosion was the trigger for the HMDF, and later (1947) opined that frequent earthquake shocks preceded the volcanic activity. Pierce (1963b:1234) surmised:

> *"The shaking motion of innumerable earthquakes, combined with the constant force of gravity, might have caused the detached fault blocks to move great horizontal distances on a slope of a few degrees."*

adding (1973:468):

> *"The common association of volcanism and earthquakes, and the close association in time between the extrusion of the Wapiti volcanics and the faulting, point up the likelihood of there having been earthquakes at that time ... The cataclysmic nature of the Heart Mountain fault movement, which is suggested by the short time involved in emplacement, is consistent with an earthquake-associated mechanism."*

Recognition of an earthquake or earthquakes as the triggering mechanism for the HMDF (our SSADF) has fallen into disfavor with most recent workers and has been replaced with a variety of ill-defined volcanic processes; *e.g.*, "collapse of a volcanic edifice" in the northern Absaroka Range (Malone and Craddock, 2008); Eocene collapse of an active Absaroka volcanic pile (Beutner and Hauge, 2009); intrusion of a lamprophyre diatreme (Malone *et al.*, 2017); and gravity-induced spreading of a continuous allochthon on the flanks of an active volcanic field (Hauge, 1985, 1990b).

Our research indicates that a megathrust earthquake or earthquakes, similar in duration and intensity to the four strongest earthquakes ever recorded by instruments, instigated the catastrophic allochthon detachment, its succeeding movement, and the immediately following volcanism. Those four earthquakes, dating since 1960, were all megathrust earthquakes that occurred along subduction zones and had moment magnitudes (Mw) of 9.1 or greater and durations of up to and exceeding 10 minutes. Length along which slip distance was estimated ranged from 500 km (311 miles) to 1,300 km (807 miles), slip width of 180 km (112 miles) to 250 km (155 miles), and rupture velocities from 2.1 km/sec (1.3 miles/sec) to 3.5 km/sec (2.17 miles/sec).

The 1960 Valdiva—Great Chilean megathrust earthquake—the most powerful earthquake ever recorded—occurred along the Peru-Chile Trench where the Nazca Plate was subducted beneath the South American Plate and had a Mw of 9.4-9.6 lasting for 10+ minutes (Barrientos and Ward, 1990). The Alaskan Good Friday megathrust earthquake of 1964 was located along the plate boundary between the subducting Pacific Plate and the North American Plate. It lasted for four minutes and 38 seconds (Stover and Coffman, 1993) and had a Mw of 9.2. The 2004 Sumatran Andaman megathrust earthquake (Lay *et al.*, 2005) had a total duration of 10 minutes in two phases and a MW of 9.1-9.3 where the India-Australian Plate is subducting beneath the Burma Plate (Sataki and Atwater, 2007), a microplate extension of the Greater Eurasian Plate. The most powerful earthquake in Japanese history, the 2011 Tohoku megathrust earthquake (Bosai, 2011), had a Mw of 9.1 and a duration of six minutes, and occurred where the Pacific Plate is subducting beneath the Okhotsk Plate, an extension of the North American Plate. Bletery *et al.* (2016a, 2016b) noted that the largest megathrust earthquakes occur where the subducting plate is either flat or has a very shallow dip, hence the term "flat slab subduction." The extremely rapid velocities and immense slip areas associated with such megathrust earthquakes cause intense and widespread shaking.

Earthquakes shake in three directions, two of them horizontal, defined as E-W and N-S, and one vertical, defined as U-D (up-down). Shaking is defined in terms of Peak Ground Acceleration (PGA), a percentage or multiple of the gravitational constant, g. The significance of the vertical component has, quite possibly, been overlooked or even neglected. With respect to the HMDF, Pierce (1973:468) observed:

> "If, during the upward cycles of oscillatory seismic motion, accelerations approaching 1g were imparted to the rocks above the incipient fault, then, at the moments that the upward accelerations ceased, the stress normal to the fault as a result of gravity would approach zero and ... in effect the upper plate would be almost weightless. With innumerable repetitions of upward acceleration, the upper plate would be intermittently nearly unrestrained by friction and free to move laterally on a very low slope."

Pierce (1973) referred to three studies of the 1971 San Fernando 6.6 Mw earthquake by Maley and Cloud (1971), Degenkolb (1971), and Morrill (1971) that, at the time, reported the highest earthquake accelerations ever recorded. Maley and Cloud (1971) provided the largest collection of strong-motion data ever obtained, in the 0.5 to 0.75 g range, with several high frequency peaks reaching 1 g. Degenkolb (1971) reported that the same record indicated ground accelerations of more than 100 percent g horizontal and 70 percent g vertical.

The point that Pierce (1973) made is that if an earthquake of moderate intensity that lasted for 12 seconds can produce recorded peak accelerations of a much higher g than expected, perhaps an earthquake of considerably greater magnitude could produce a peak vertical acceleration approaching or exceeding 1 g, which would cause loss of cohesion (friction) and, combined with gravity, would be sufficient to initiate movement on a gently dipping surface.

Data from the 2011 Japanese Tohoku megathrust earthquake indicates a horizontal acceleration of 2.7 g and a maximum upward vertical acceleration of 1.88 g (National Research Institute for Earth Science and Disaster Prevention). The 1897 Great Assam earthquake in India, an estimated 8.2-8.3 Mw megathrust earthquake, exhibited vertical acceleration exceeding 1 g (Bilham and England, 2001). Korzec (2016) performed analyses on a large number of acceleration time histories revealing that vertical peak ground acceleration can be as high as horizontal acceleration, and Bozorgnia and Campbell (2016) concluded that, for shallow crustal earthquakes in active tectonic regions with MW ranging from 3.3 to 8.5 and a number of fault parameters, the V/H ratio ranged from about 0.39 to 0.60.

Thus, Pierce's (1973) conjecture appears to be valid.

Megathrust earthquakes of the intensities and durations discussed above are relatively infrequent, perhaps three or four occurring every 100 years (Hayes et al., 2017) but, over the course of geologic time, even at one/century, up to two million or more such earthquakes may have occurred since the breakup of Pangea. Given this hypothetical frequency of great megathrust earthquakes, it seems likely that one of these taking place under the unusual geologic conditions at the opening of an immense volcanic field might well have initiated the SSADF.

Farallon Plate rollback beneath western North America during the Eocene (Smith et al., 2014; Currie and Copeland, 2022) provides a mechanism for the volcanism associated with the SSADF. Figure 2 in Smith et al. (2014) presents a series of paleogeographic time slice maps dated from 53 to 47 Ma with interpreted curved or arcuate axes which define active periods of volcanism in the Absaroka

Volcanic Province. Those arcs moved essentially north-to-south across northwestern Wyoming in response to rollback. Equivalent rocks, oldest to youngest the Washburn, Sunlight, and Thorofare Creek groups developed progressively from north to south. Between 50.0 Ma and 48.5 Ma the maps illustrate active magmatism in northwestern Wyoming and adjacent Montana. Smith *et al.* (2014, Figure 4) show a series of cross-sections depicting the progressive rollback of the Farallon Plate from 53 to 47 Ma and the upwelling of hot aesthenospheric mantle that initiated volcanism in the overriding North American Plate. The Cathedral Cliffs Formation, a unit of the Washburn Group, pre-dates the SSADF detachment. The 50.0 Ma time slice map appears to correlate well with the potential timing for Cathedral Cliffs volcanism and deposition. Wapiti volcanism also correlates well with time slice maps 48.5 Ma and 47.0 Ma.

Chadwick (1970) identified 13 volcanic centers in the Absaroka-Gallatin Province, extending 165 miles (270 km) from Livingston, Montana, to Dubois, Wyoming, in two subparallel trends/belts 60 miles (100 km) wide from the Buffalo Bill Dam west of Cody, Wyoming, to Lake Butte in eastern Yellowstone National Park. The eastern belt is more potassic rich, whereas the western trend is potassium-poor and more mafic. The eastern belt contains the oldest Absaroka Volcanic Province rocks—the Washburn Group—that is partially comprised of the Cathedral Cliffs Formation. Feeley and Cosca (2003) observed that at least three eruptive centers (Sunlight, Crandall, and Independence) in the eastern belt were active contemporaneously. Sunlight Volcano erupted in three distinct major effusive stages, at about 49.4 Ma, 48.4 Ma, and 48.1 Ma. They also reference several studies indicating that the two subparallel belts appear to be localized along Precambrian zones of crustal weakness that were reactivated during the Middle Mesozoic to Early Tertiary Laramide Orogeny. Smith *et al.* (2014, Figure 2) illustrate that at 50.0 Ma and 48.5 Ma the arcuate magmatic belt covered northwestern Wyoming and adjacent Montana and included the 13 eruptive centers described by Chadwick (1970). Even though Feeley and Cosca (2003) mention only three of the active volcanic centers, and Sunlight specifically, the other ten volcanic centers must have been active at some time during the 53-47 Ma time span to have contributed the immense cumulative volume of ~30,000 km^3 of volcanic rocks distributed over an area exceeding 23,000 km^2 (Smedes and Prostka, 1972). Although Feeley and Cosca (2003) argue against mantle involvement in the eruptions of Sunlight Volcano, we suggest that the presence of olivine, pyroxene, and basalt in the Wapiti volcanics and their absence in the Cathedral Cliffs Formation supports more mantle involvement in Wapiti volcanism.

An alternative interpretation of Absaroka volcanism from 53-47 Ma can be summarized from Haeussler *et al.* (2003) who provided a map defining Absaroka volcanics as resulting from 53-47 Ma Eocene "back arc" magmatism, and from Madsen *et al.* (2006) who depicted Absaroka volcanics to be "Eocene extended arc to back arc magmatism". The Absaroka volcanics in northwestern Wyoming and adjacent Montana are the southeastern limit of the Challis-Kamloops Belt that extends into northern British Columbia and was active from 53-45 Ma. Madsen *et al.* (2006) contended that a Farallon Plate-Resurrection Plate slab window extended from below southern British Columbia into northwestern Wyoming. This slab window is roughly coincident with the Challis-Kamloops Belt.

Our studies suggest the following sequence of events best explains emplacement of the HMDF (SSADF): 1) An initial period of volcanism at about 49.5 Ma; 2) A megathrust earthquake probably exceeding Mw 9.0 that occurred during flat slab subduction of the Farallon Plate beneath the North American Plate triggered the subhorizontal rupture and detachment within the Bighorn Dolomite; and 3) Immediately subsequent volcanism. The first period of volcanism was caused by upwelling of hot aesthenospheric mantle during Farallon Plate rollback (or, alternatively, through a slab window between the Farallon and Resurrection plates) that melted the locally overlying continental lithosphere of the North American Plate. Deposition of the Cathedral Cliffs Formation and equivalents on an eroded surface of Paleozoic carbonate rocks was the result of this volcanic event. The seismic activity of this earthquake lasted several minutes and induced the catastrophic "breakaway" of an upper plate carapace consisting of 460-610+ m (~1,500' – 2,000'+) of largely Ordovician through Mississippian carbonate rocks and the Eocene volcanics from the first period of volcanism (Cathedral Cliffs Formation), and its rapid (160-1,224 km/hr = 100–760 mph) transport to the southeast down an increasingly tectonically denuded slope of 1°-2°. This surface of tectonic denudation in the basal Bighorn Dolomite (**Figure 16**) became stripped and overridden by an advancing, continuous allochthon that fragmented into individual blocks. The rapid downslope advance of the leading edges of the allochthons slowed at the rise at the base of the transgressive ramp at the eastern edge of the Sunlight Basin, but the momentum of some of the allochthons propelled them up the ramp into the area between Pat O'Hara Mountain and Dead Indian Hill. The larger blocks that remained in the eastern Sunlight Basin have quite possibly been eroded into smaller masses by the present-day drainage system.

The SSADF megathrust earthquake probably preceded the eruptions of Sunlight Volcano, and that earthquake was then followed by the immediate subsequent volcanism of Sunlight Volcano during the continued rollback of the Farallon Plate (or through the slab window). This volcanism resulted in rapid deposition of the Wapiti Formation which buried both the detached blocks and the surface of tectonic denudation.

The post-breakaway basal lava of the Jim Mountain Member of the Wapiti Formation was deposited at 49.5 +/- 0.16 Ma (Feeley and Cosca, 2003), firmly placing both the SSADF and the beginning of Wapiti deposition in the later early Eocene (early Eocene = 55.93 – 47.80 Ma; Westerhold *et al.*, 2018). Craddock *et al.* (2009) noted that the Crandall Volcano is the volcanic locus most proximal to the breakaway area of the HMDF (SSADF), and those authors suggested that movement on the detachment fault occurred between 49.7 and 47.5 Ma.

In discussing earthquake activity as having had a role in producing the initial rupture, Pierce (1963b:1234-1235) noted that:

> *"... some additional factor, as yet undetermined, may have been required to produce the initial rupture along the bedding and transgressive faults."*

The 1°-2° east to southeast regional dip of Paleozoic rocks was probably the result of Laramide orogenic processes prior to the detachment. Alternatively, this tilt may have accompanied doming of surface strata from dynamic and thermal uplift under volcanic centers during a Farallon Plate rollback as proposed by Smith *et al.* (2014). This dynamic and thermal uplift was caused by upwelling aesthenosphere that preceded volcanism. Storage of the upwelling magma may have occurred in both deep and shallow chambers similar to Anak Krakatau Volcano (Dahren *et al.* 2012). The deep chambers would have been at or near the base of the crust, whereas shallower chambers were situated much closer to the surface at depths of 1-10 km (0.6-6.0 miles). The filling of magma chambers may uplift the land surface by 10s to 100s of meters. This uplift is known as *precaldera tumescence* and, combined with elastic deformation, causes structural doming (Branney and Acocella, 2015). Magma chambers lying deeper may cause less uplift but may affect a wider region, whereas shallow magma reservoirs may readily raise the overlying crust as they inflate (possibly resulting in the "inflation of the flank" of the Crandall Volcano of Siebert *et al.*, 1987). The dynamic and thermal uplift that resulted from filling of shallow magma chambers under any one or more of the volcanic centers may have provided lateral pressure sufficient to instantaneously rupture the Bighorn Dolomite all along a basal horizontal plane of detachment (see also Straw and Schmidt, 1981). Alternatively, the massive megathrust earthquake discussed earlier may have resulted in the sudden separation of stressed strata along a gently curved horizontal plane, much like the parting of two foils of an onion.

The allochthon/autochthon contact being recognized everywhere at the same level near the base of the Bighorn Dolomite is incontrovertible evidence that the initial movement following the breakaway was that of a single, massive sheet.

Siebert (1984) suggested that the intrusion of swarms of parallel dikes (such as those invading some SSADF allochthons) might have contributed a dilational effect, though we deem this explanation unlikely. Equally improbable is the contention by Malone *et al.* (2017) that initiation of the breakaway and the initial movement of the HMDF (our SSADF) was caused by a lamprophyre diatreme eruption somewhere to the west of the present location of White Mountain. The lamprophyre intruded carbonate rocks that had been previously marbleized by a diorite stock (**Figures 16** and **17**). The stock and a remnant of the lamprophyre dike are now preserved at White Mountain. Malone *et al.* (2017) believe that White Mountain is a remnant of the eruptive center. Given the small remnant of the diatreme (**Figure 17**), it seems highly unlikely that this particular single eruption had any appreciable role in SSADF breakaway and/or displacement.

A preponderance of evidence from upper plate/lower plate contact relationships indicates that at least the earliest stages of post-breakaway movement were facilitated by one or more lubricating mechanisms, and that modeling of these myriad agencies points overwhelmingly to extremely rapid, catastrophic movement (Beutner and Craven, 1996; Beutner and Gerbi, 2005; Craddock *et al.*, 2009; Anders *et al.*, 2010; Goren *et al.*, 2010). Suggested mechanisms include: a volcanic explosion (Bucher, 1933); gas lubrication (Bucher, 1933; Hughes, 1970); fluid pressure, fluid flow, and general fluidization (Hubbert and Rubey, 1959; Templeton *et al.*, 1995; Douglas *et al.*, 2003; Anders *et al.*, 2010); fluid-wedge pressure (Voight, 1972, 1973c); water vaporization induced by creep (Goguel, 1969); "phraeto-magmatic hydraulics" (Straw and Schmidt, 1981); rock glacier development (Sales, 1983); "hovercraft" on compressed volcanic gas (Hughes, 1970) or air (Shreve, 1968); acoustic fluidization (Melosh, 1983, 1986, 2018); clastic fluidization (Voight, 1973a); fluid flotation (Voight, 1973b); pore fluid vaporization (Goguel, 1978); air cushion (Hsü, 1969); a viscous zone of décollement (Kehle, 1970); volcanic fluidization (Beutner and Craven, 1996—but see Colgan, 1998); accreted grains (Beutner, 2002); hot water (Aharonov and Anders, 2008); heated pore fluids (King *et al.*, 2009); thermal decomposition of carbonates (Goren *et al.*, 2010; Mitchell *et al.*, 2015); and "episodic dissolution" (Swanson *et al.*, 2016).

Although Davis (1965) criticized the role of fluid pressure in HMDF (SSADF) movement, the breakaway having at the outset involved an immense, sheet-like allochthon of Paleozoic rocks, it seems probable, even likely, that one or more of the aforementioned processes operated, at least locally, for a very brief time. No one of the above hypothetical mechanisms has enjoyed an especially wide currency at the expense of the others, and none constitutes a prerequisite for initiating or sustaining movement under conditions of continuous seismic activity, as proposed by Bucher (1947) and argued for by Pierce (1973). The majority of the above listed contact phenomena studies support or are consistent with rapid, catastrophic movement of the HMDF (SSADF).

Losh (2018) observed that movement-generated CO_2 gas cannot also have initiated movement which, ironically, he attributes to a volcanic tremor—a seismic mechanism he inexplicably discards as incapable of sustaining movement. Straw and Schmidt (1981) also require earthquake motion to initiate movement. The same reasoning applies to all of the other physico-chemical and hydraulic solutions for facilitating and/or sustaining movement of the detached mass.

Of Pierce's (1957) model of catastrophic emplacement of the HMDF allochthons and the formation of a surface of tectonic denudation, Hauge (1990b:1174) observed that:

> "The mechanics of the process of catastrophic emplacement of the slide blocks envisioned in this model remains enigmatic."

The perceived enigma (also a criticism of the tectonic denudation model voiced by Voight [1974] and Hauge [1993]), obtains from the perceived lack of a clear mechanism by which the HMDF attained tens of kilometers of runout on a 2° slope (Goren et al., 2010). We contend that the clear mechanism for breakaway, deployment, and emplacement is that of sustained seismic activity proposed by Pierce (1979), and that process alone, in the context of megathrust earthquakes, was sufficient to reduce friction between the autochthon and allochthon and to both initiate and sustain several minutes of rapid, catastrophic movement resulting in a massive displacement. In any case, we agree with Pierce (1963a) that, once the upper plate fractionated into several blocks, sufficient confined fluid pressure to reduce friction (of whatever origin) could not be maintained. In instances in which brecciated or comminuted Paleozoic carbonate debris underlay allochthons, however, such debris would have lessened friction and facilitated movement and, in the absence of gases or liquids, any gouge debris/ breccia produced would be less likely to be forced out as the blocks spread apart. Rengers (1970) concluded that the amount of friction between separation planes is a factor of the roughness of the planes (i.e., in the instance of this study, the planes being those at the top of the autochthon and the base of the allochthon). We contend that, together, the immense area of the initial allochthon and the confined stratigraphic interval of the plane of separation render the degree of roughness minimal and the amount of friction between the SSADF plates inconsequential.

Rempel and Rice (2006) demonstrated that heating can be rapid at the bases of large landslides, and Mitchell et al. (2015) showed that the thermal decomposition temperature of dolomite (>600° C) can be reached with only a few hundred μm of movement. Fluid pressure could be generated and maintained until the breakup of the mobile carapace into several allochthons. After that, the reduction of friction and sustaining of motion could well be facilitated by exceptionally violent megathrust seismic activity.

In sum, we propose that a megathrust earthquake (on the order of Mw 9.0 + lasting for several minutes) resulted in the breakaway of an upper plate carapace of Paleozoic carbonate rocks and pre-detachment Eocene volcanics, catastrophically propelling the upper plate to the east and southeast, the separation of the carapace into individual blocks, scattering of the blocks down dip, and their ultimate emplacement. Intense volcanism prior to and immediately subsequent to the breakaway resulted from subduction of the Farallon Plate beneath the North American Plate.

To recapitulate, megathrust earthquakes are the only kind known to produce Mw 9.0+ earthquakes. Four of the most powerful of these earthquakes in the modern era are: 1) the 1960 Valdivia (Chile) megathrust earthquake (e.g., Barrientos and Ward, 1990) which was measured at Mw 9.4-9.6 and was sustained for 10+ minutes; 2) the 1964 Alaska megathrust earthquake (e.g., Stover and Coffman, 1993) measured Mw 9.2 and had a duration of 4 minutes and 38 seconds; 3) the 2004 Sumatran Andaman megathrust earthquake (Lay et al., 2005) measured Mw 9.1-9.3 and lasted for 10 minutes in two phases; and 4) the 2011 Japanese Tohoku megathrust earthquake (Bosai, 2011) had an Mw of 9.1 and a duration of six minutes.

It is logical that the time of initial volcanism, followed immediately by the megathrust earthquake and continuous high Mw seismicity, would coincide with the time of origin of Sunlight Volcano; *i.e.*, immediately preceding lower Wapiti deposition (at about 49.44 Ma; Malone, Schroeder, and Craddock, 2014) and immediately succeeding deposition of the Crandall Conglomerate (at about 49.57 Ma; Malone *et al.*, 2015), or about 49.5 Ma. We therefore are in agreement with Pierce (*e.g.*, 1973) and Voight (1974) that major volcanic activity immediately preceded the SSADF and immediately followed cessation of movement.

Distance of Movement (Length of Run-out) of Allochthonous Rocks

In his recapitulation of the "Heart Mountain Fault and the Mechanism Problem," Pierce (1973) attributed displacement distances of 35 miles (57.4 km) to the bedding plane part of the fault, 9 miles (14.8 km) to the transgressive part of the fault, and 30 miles (49.2 km) to that part of the fault where the allochthon traversed the former land surface. It is easy to misconstrue those figures to suggest that displacement along the Heart Mountain Detachment Fault is the sum of those figures, = 74 miles (121.4 km); *i.e.*, that the allochthon moved 74 miles from its source at the Abiathar Peak "breakaway" fault (Pierce, 1960; Prostka *et al.*, 1965) to the farthest-flung remnants on McCullough Peaks. For example, it is obvious that allochthonous rocks near Silvergate, MT and Cooke City, MT (nearest the "breakaway"; *e.g.*, at Republic Mountain) were displaced at most several hundred meters to a few kilometers.

Previous estimates of the HMDF run-out have varied considerably: 22 miles (=36 km; Dake, 1918); 28 miles (=46 km; Aharnov and Anders, 2006; Anders *et al.*, 2010); 31 miles (=51 km; Hauge, 1993; Goren *et al.*, 2010; Losh, 2018); 53 miles (=85 km; Malone, Craddock, and Mathesin, 2014); 66 miles (=108 km; Craddock *et al.*, 2012). What is not revealed by those authors regarding displacement distance is how those distances were determined; *i.e.*, where did those authors believe movement of specific elements of the allochthon began and how distant from source did those specific elements stop?

Subtracting the allochthonous rocks we now attribute to the Heart Mountain Sturzstrom deposit (HMMPS, see below) from Pierce's (1973) estimate of the run-out of the HMDF, we are still left with an enormous stream of immense, SSADF allochthons distributed over a considerable area spanning, in the long dimension, the 36-mile (59 km) distance between Abiathar Peak and the Dead Indian Summit. But how many of those 36 miles constitute the true run-out of the HMDF (= the Shoshone/Sunlight/Abiathar Detachment Fault)? In other words, did the Paleozoic allochthons on the former land surface above the ramp near the Dead Indian Summit really originate on the breakaway scarp near Abiathar Peak? Voight (1974, his Figure 13) provided a series of schematic maps illustrating a sequence of events in which allochthons closest to the initial breakaway remain near that breakaway and allochthons increasingly farther away were detached from the advancing front. We endorse a similar interpretation with a maximum 15.3 miles (~25 km) of movement.

We suggest that that the straight-line distance of 36 miles *is simply the distance over which rocks affected by HMDF=SSADF faulting are distributed—it is no measure of displacement or run-out distances; that is, it does not measure how far the allochthonous rocks moved from source.* Perhaps it is most accurate to say that movement of some kind, of a whole and/or fragmented allochthon, took place over a vast area having a northwest-southeast long axis measuring about 36 miles (~59 km) in length. Whereas allochthonous masses closest to the breakaway escarpment almost certainly had their sources somewhere along that scarp, masses occurring farthest away from that scarp (*i.e.*, those at the Dead Indian Summit and at Sheep Mountain) were almost certainly derived from the breakup of the advancing southeastern front of the upper plate (Pierce, 1957:617). Under that construct few, if any, of the allochthonous Paleozoic blocks moved farther than about 25 km.

Buetner and DiBenetto (2003) determined that the upper plate and lower plate distributions of the Crandall Conglomerate (Pierce and Nelson, 1973) indicate that unit was displaced 25-30 km by movement along the HMDF (=SSADF). We propose that measurement of the displacement of fault-truncated bodies of the Crandall Conglomerate is perhaps the only empirical manner by which the maximum amount of displacement on part of the SSADF might be estimated. It is a reasonable assumption that one or more of Pierce and Nelson's (1973, Figure 7) outcrops of transported, upper plate bodies of Crandall Conglomerate (**Figure 18**) correlate(s) with one or more of their outcrops of truncated bodies of the Crandall Conglomerate on the autochthon (**Figure 19**). From their mapped data, we determined a distance of approximately 15.3 miles (=24.5 km) of displacement by measuring the maximum separation of the westernmost lower plate exposure of the Crandall Conglomerate and its westernmost upper plate exposure, and the maximum separation of the easternmost lower plate exposure of the Crandall Conglomerate and its easternmost upper plate exposure 14.9 miles (=24 km). To the extent that the upper and lower plate exposures may be so correlated, there is no other empirical field evidence supporting any lateral displacement on the SSADF exceeding 15.3 miles (24.5 km).

The allegation by Hauge (1990a, 1990b) that the distribution of Paleozoic rocks of the allochthon might be due to erosion prior to Wapiti deposition is obviated by the many occurrences of the truncated/transported Crandall Conglomerate on the lower and upper plates, respectively, of the detachment fault.

Speed of Movement

Emplacement of SSADF allochthons was probably complete a few minutes after movement commenced (*e.g.*, Beutner and Gerbi, 2005). Individual allochthons are largely intact, a circumstance that suggests to us a single, very rapid episode of movement. Repeated instigation of movement of such massive, inert blocks would require repeated violent seismic events and would result in intense deformation, loss of internal cohesion, and disintegration of at least some of the blocks. Rapid, sustained movement with minimal friction at the allochthon/autochthon contact could be maintained by continuous seismicity, as originally suggested by Bucher (1947) and long maintained by Pierce (1963a, 1973). Voight (1974:28) referred to the Bucher/Pierce interpretation of earthquake oscillation-induced movement of the HMDF (=SSADF) allochthons as a "vibrating conveyor," a mechanism nearly identical to that produced by the vibrating football game table sold by Tudor[tm], in which the football "players" are advanced down the field by electronically induced vibration of the metal playing field. With the SSADF, several minutes of sustained seismic activity (vibration) fed energy continuously into the detachment fault system and kept the blocks moving. Under the Hauge interpretation, however, inertial friction would remain maximal, inertia would need to be continuously overcome over an immense area for 1-2 million years, and deformation at the upper plate/lower plate interface would, we contend, have been intense.

Craddock *et al.* (2009) estimated that detachment fault movement lasted 3-4 minutes. Goren *et al.* (2010) suggested speeds of 10s of meters/second up to 100 meters/second (360 km/hr), and Losh (2018) predicted an emplacement speed of 10s of miles/hour. Other authors offered more precise estimates of emplacement speed: 93 mph (=150 km/hr; Beutner and Gerbi, 2005); 100 mph (=161 km/hr; Losh, 2018); 115 mph (=185 km/hr; Voight, 1973b); up to 100 m/second (=224 mph = 350 km/hr; Goren *et al.*, 2010); >100 m/second; (>224 mph = >350 km/hr; Craddock *et al.*, 2009); 150 m/second and 126-340 m/second (=335 mph = 540 km/hr and 282-760 mph = 454-1,224 km/hr; Malone and Craddock, 2008; Craddock *et al.*, 2009).

The time of emplacement over a distance of 24.5 km (~15 miles) is 1.2 minutes at 1,224 km/hr (760 mph), 4.2 minutes at 360 km/he (224 mph), and about 10.0 minutes at 150 km/hr (93 mph). These time estimates for emplacement of all parts of the upper plate are quite reasonable as they lie well within or are less than the measured time range of sustained earthquake activity documented for the four greatest historic megathrust earthquakes: the Alaska megathrust earthquake (about 4.6 minutes), the Valdiva megathrust earthquake (in Chile; about 10 minutes), the Sumatran Andaman megathrust earthquake (in Indonesia, 10 minutes in two phases), and the Japanese Tohoku megathrust earthquake (6 minutes).

✥ Analogous Structure

Wolfe (1977) recorded 5 km (3.125 miles) of displacement of an immense block of Miocene limestone down a low-angle (0.6°) slope during late Holocene time on Samar Island in the Philippine Republic. The block became isolated by normal faulting and, movement of the block, which measures approximately 18 km X 25 km (11.25 miles X 15.625 miles) by about 400 m in thickness (= 180 km^3 or 43.68 miles^3), produced a "strip plain" analogous to Pierce's surface of tectonic denudation. The displacement is presumed to have been triggered by the fluidization of an underclay by earthquake activity.

The Philippine Archipelago, known structurally as the Philippine Mobile Belt, is an island arc system and one of the most seismically active regions on earth (Yumul *et al.*, 2008). Samar Island lies between the Philippine Trench to the east and the Philippine Fault Zone to the west. The west-dipping Philippine Sea Plate is being obliquely (northwestwardly) subducted beneath the Philippine Archipelago at the Philippine Trench, whereas the Philippine Fault Zone is a left-lateral NW-SE-trending strike slip fault that bisects the archipelago. The archipelago itself is essentially bounded on the west by the Manila, Negros, Catabato, and Sula trenches, where east-dipping subduction of the Eurasian and India-Australian plates is taking place. It is easy to envision frequent and intense earthquakes and volcanism in this highly active tectonic setting.

Although the Samar Island allochthonous mass is considerably smaller than the originally intact Shoshone/Sunlight/Abiathar Detachment Fault (SSADF) allochthon, its volume exceeds that of any individual SSADF block, and it is the only other known gravity-induced detachment structure in which a comparable "surface of tectonic denudation" has been preserved.

HEART MOUNTAIN/McCULLOUGH PEAKS STURZSTROM (HMMPS)
(T.M. Bown, A.J. Warner, and M.E. Mathison)

Introduction

Long ignored in the spectrum of remarkable geological phenomena attributed to the Heart Mountain Detachment Fault (HMDF) are Heart Mountain itself—the namesake landmark of the phenomenon—and the allochthonous Paleozoic carbonate debris atop (Hewett, 1920) and proximal to (Sinclair and Granger, 1912; Bown and Love, 1989) McCullough Peaks—masses heralded as the most distally emplaced remnants of the "long runout" of the detachment fault. What Eldridge (1894) and Fisher (1906:37) originally termed a "circular fault" at Heart Mountain (= "Hart Mountain" of Dake, 1918) in recent years has been expanded in areal extent considerably to the west, east, and south by succeeding authors, most notably by Malone *et al.* (2014) who included allochthonous volcanic rocks atop the 50 miles (82 km) distant Squaw Buttes in their much-expanded concept of the Heart Mountain Detachment Fault.

Distribution and Composition of Debris

Three geographically distinct areas exhibiting large volumes of allochthonous Paleozoic carbonate rocks occur in the northwest Bighorn Basin, east of and extraneous to the front of the Beartooth Range/Dead Indian Hill/Pat O'Hara Mountain/Rattlesnake Mountain massif. These are: (1) the jumbled blocks of Ordovician (Bighorn Dolomite), Devonian (Jefferson Limestone and Three Forks Formation limestone and shale), and Mississippian (Madison Limestone) carbonate rocks atop Heart Mountain, north of Cody, WY (**Figures 20 and 21**); (2) remnants of a blanket of largely comminuted Paleozoic carbonate debris supporting atop them a few tabular masses of shattered dolomite or micrite breccia up to several tens of meters in diameter and hundreds of individual blocks of lesser size, the debris attaining up to 85 meters in thickness capping parts of McCullough Peaks (**Figures 22-24**); and (3) remnants of largely brecciated Paleozoic carbonate debris distributed across parts of the rugged badland slopes of the western margin of the McCullough Peaks eminence, about 1.2-4.0 miles (about 1.9-6.4 km) east and northeast of Corbett Dam (Sinclair and Granger, 1912; Bown and Love, 1989). Here termed the Corbett masses (**Figures 25-28**; Table II), these last debris masses were lumped with "carbonate pediment deposits" by Pierce (1973, his Fig. 3), and lie approximately 200-500 meters (656-1,640 feet) above the adjacent Shoshone River, and about 100-400 meters (328-1,312 feet) beneath what Hewett (1920) termed the "Middle Peak" (the highest point) of McCullough Peaks (**Figure 29**).

We define the composition of HMMPS debris in five categories, distributed as follows:

1) Immense, intact, stratified blocks (Heart Mountain, **Figures 20 and 21**);

2) Masses of dolomite and/or micrite breccia cemented with authigenic quartz probably derived from chert nodules in the carbonates (Corbett masses and principal masses of Middle and East Peaks of McCullough Peaks, **Figures 22, 23, 25-28, and 30-32**);

3) Comminuted, unsorted, uncemented carbonate debris varying in size from rock flour to boulders in size (atop pediment surfaces proximal to Heart Mountain, debris in Rattlesnake Canyon, Figure 33);

4) Boulder to massive block-size carbonate clasts (pediment surfaces distal to Heart Mountain, and distal part of McCullough Peaks pediment, Figures 34 and 35);

5) Veneers of uncemented, finer-grained, thumb- to fist-size carbonate rubble (distal parts of McCullough Peaks and Heart Mountain pediments, Figures 35 and 36).

Heart Mountain Allochthons

Bucher (1947, his Fig. 1) depicted the McCullough Peaks masses as "Madison," and those on Heart Mountain as "Bighorn," whereas it is now clear that allochthonous rocks of both units, the Mississippian Madison Limestone and the Ordovician Bighorn Dolomite, occur in both areas. Rocks of the Jefferson and Three Forks formations, of Devonian age, are also locally present. The diagrammatic cross-section of Heart Mountain by Stevens (1938, Plate 5) is an accurate illustration of the distribution of allochthonous Paleozoic rocks there, and depicts the western and eastern ends of the mountain to be made up largely of mounds of shattered and disaggregated Bighorn Dolomite (Ordovician) capped by peaks of the Devonian Jefferson Formation and separated by a titanic, monolithic, tilted central block of Mississippian Madison Limestone (Figures 20 and 21). No volcanic or volcaniclastic material occurs anywhere on Heart Mountain, and no parts of the allochthonous Paleozoic carbonate rocks there evince any sign of having been buried by volcanic rocks; *i.e.*, there has been no cementation of comminuted carbonate rock by chemicals or fine sediments of volcanic origin, and there is no staining, contact metamorphism, or other alteration of those carbonates through reaction to burial by volcanic material.

Pierce (1966, 1997) mapped but a quite restricted area of Paleozoic rocks on Heart Mountain as allochthonous, including all the immediately surrounding area in the lower Eocene Willwood Formation. Perhaps he believed that the up to 30 m of comminuted Paleozoic dolomitic and micritic debris (mixed, unsorted rock flour to angular blocks up to 3+ m in diameter), that mantles the Willwood Formation over a much larger surrounding area and caps several outlying pediment remnants, to have weathered and fallen from the massive allochthonous blocks. That material is actually the residue of an enormous volume of landslide (sturzstrom) debris that was emplaced at the same time as the massive blocks on the upper reaches of the mountain.

Paleozoic debris on Heart Mountain originally engulfed an enormous pediment surface, several erosional remnants of which are preserved to the north (Figure 37)(*e.g.*, at 44° 42' 29" N, 109° 08' 56" W, 44° 42' 10" N, 109° 05' 58" W, and 44° 40' 30" N, 109° 08' 13" W), whereas landslide debris is evident at least as far north as 44° 45' 25" N, 109° 04' 33" W (*e.g.*, near the area of Figure 34), and to the south (*e.g.*, at 44° 39' 36" N, 109° 06' 17" W) of Heart Mountain. Tracing extensions of the loci of the upward-sloping surfaces of these erosional remnants demonstrates that they were once confluent with the surface upon which the displaced Paleozoic carbonate masses forming the upper reaches of Heart Mountain lie. The southern pediment remnant is especially well preserved in this regard, and sighting up the loci of the tops of the pediment surface remnants at a bearing of about 345° from the vantage point at 44° 36' 09.3" N, 109° 06' 10.6" W exhibits this confluence quite clearly (Figures 38-40). This southern pediment remnant is mantled with a shallow veneer of rounded quartzite pebbles and cobbles and angular chert pebbles. Stevens (1938:1260) noted the like occurrence of quartzite "pebbles and cobbles" weathering out of the "greatly contorted" Wasatch (= Willwood) Formation rocks outcropping directly beneath displaced Paleozoic rocks on the north side of Heart Mountain. Field relationships demonstrate those clasts were derived from pediment mantle deposits overridden by HMMPS landslide debris.

Were there other blocks of Paleozoic carbonate debris of Heart Mountain or similar size displaced into the western part of the Bighorn Basin that have been eroded away? Probably not. Considering the distribution of Paleozoic allochthonous material and the geomorphology and topography of the Heart Mountain/McCullough Peaks region, it is likely that the high erosional remnants of Heart Mountain itself and the highest topographic levels of McCullough Peaks and intervening areas probably owe their elevated positions exclusively to the resistant carbonate debris that once mantled them or mantles them today and has protected them from erosion.

McCullough Peaks Allochthons

McCullough Peaks is a high eminence (maximum elevation = 6547 feet = 1996 m) rising above the Shoshone River valley about 15 miles (= ~24 km) at a bearing of about 115° from Heart Mountain. Hewett (1920) was the first to describe blocks of Paleozoic carbonate rocks on top of the Peaks, where the greatest thickness of transported Paleozoic debris is the approximately 85 meters (about 279 feet) of largely comminuted limestone and dolomite breccia making up East Peak (**Figures 22 and 29**) . The designations used here of the three highest elevations on McCullough Peaks as West, Middle, and East Peaks follows the usage of Hewett (1920:547) who, later in his 1920 text (p. 551), quixotically changed the name East Peak to South Peak for the most southeastern of the three peaks—that located at 44° 34' 29.41" N, 108° 49' 34.93" W. We continue the usage of "East Peak." West Peak (**Figure 29**) is also known locally as "Pyramid Mountain."

Hewett (1920:551) described a "triangular cap" of Madison limestone about 600 feet x 400 feet x 80 feet thick (= 19,200,000 ft^3 = 57,682 m^3) at East Peak (his South Peak); however, we were unable to determine whether that outcrop was of a single mass or of several juxtaposed blocks. The largest single intact mass we located on McCullough Peaks is another roughly triangular, shattered block of Bighorn Dolomite situated about 750 meters southeast of East Peak, at 44° 34' 20" N, 108° 49' 04" W (**Figures 23 and 32**). With a base length of about 48 meters and a height of about 102 m, the area of the triangle circumscribing the block is about 2,450 m^2 or ~1/4 hectare. The thickness of the block is difficult to ascertain but appears to average about 7 meters, so the volume of the largest transported intact block of Paleozoic carbonate rock atop McCullough Peaks is approximately 17,150 m^3.

Dozens of smaller hilltop remnants of comminuted carbonate rock debris and thousands of blocks of limestone and dolomite of varying sizes lie scattered upon the upper levels of the McCullough Peaks rise (**Figure 24**), and these were distributed across a surface of considerable relief. Although it is clear that a large volume of the debris, including much of that lying southeast of East Peak, may owe its current position to sheetwash (*e.g.*, Abrahams *et al.*, 1984) and surface creep (*e.g.*, Leopold *et al.*, 1966; Schumm, 1967; Radbruch-Hall, 1978) over the past several hundreds of millennia, the presence of numerous hillocks capped by several meters to tens of meters of *in situ* comminuted carbonate debris indicates that the mobile allochthonous material of the HMMPS encountered an irregular badland topography with a relief of about 140 meters before its ultimate passing across the (then) relatively undissected, southeast-sloping McCullough Peaks pediment.

The topography of the residual landslide debris capping McCullough Peaks consists of rounded, hummocky hills with closed basins and rafted blocks similar to the classical surface aspect of landslide debris (*e.g.*, Heim, 1883; Abele, 1994) and wholly unlike the surrounding jagged, eroded badland peaks of the underlying lower Eocene Willwood Formation or the extensive mantled pediment to the southeast. The easternmost margin of the landslide debris atop McCullough Peaks rises above and abuts the surface of a relatively flat, pre-landslide pediment surface (**Figure 22**) that descends to the southeast at an inclination of about two degrees and is irregularly surfaced with a thin veneer of carbonate debris ranging in grain size from rock flour and granular particles up to angular blocks 90 x 70 x 46 cm (~0.3 m^3) in volume. At a few localities this angular Paleozoic carbonate debris fills pre-landslide erosional scours cut up to six meters into the pediment surface (**Figure 36A**). Such irregular variations in the thickness of the debris covering the relatively flat pediment surface indicates that the distal part of the McCullough Peaks pediment surface was developed and its surface eroded before and modified after the allochthonous Paleozoic debris was deposited. Although the debris mantling the pediment surface is dominantly of Paleozoic carbonate origin, a small proportion of rounded to subrounded granules and pebbles of volcanic origin is present at most localities we visited (*e.g.*, at 44° 33' 07.39" N, 108° 45' 57.33" W—a typical site), and represents pre-landslide pediment mantle debris.

The Corbett Masses

Sinclair and Granger (1912) first recognized blocks of limestone and dolomite and masses of comminuted carbonate debris strewn across the rugged Willwood Formation badlands making up the west slope of McCullough Peaks. This material has the same composition as that on the upper levels of McCullough Peaks, and masses up to 35 meters thick are distributed irregularly across the badland area some 2.5 miles (4.15 km) northeast of Corbett Dam and lying 200-500 meters above the Shoshone River (Table II). Pierce (1973) attributed these deposits, that we term the Corbett masses, to a pediment origin; however, they differ considerably in thickness, composition, and physical aspect from pediment deposits that Pierce mapped and described on the Clark (Pierce, 1965a) and Cody (Pierce, 1966a, 1978, 1997) quadrangles.

Pediments are sloping surfaces formed by lateral corrasion (Johnson, 1885; Davis, 1930—but see also Mears, 1993), they commonly possess a veneer of alluvium (Hadley, 1967), and their development generally represents formation over extended periods of erosional stability. Remnants of pediments surrounding Heart Mountain and that forming the southeastern slope of McCullough Peaks are strewn with granules, pebbles, and boulders of largely angular debris derived mostly from allochthonous Paleozoic carbonates that were dispersed—following landslide deposition—by sheet-flooding and surface creep. Those surfaces differ both physiographically and compositionally from the Corbett masses.

In 1989 it was suggested to T.M. Bown that the Corbett masses were somehow "let down" from thicker, intact bodies of allochthonous debris once capping higher areas on McCullough Peaks. Nelson *et al.* (1972) mapped three of the masses lying between McCullough Peaks and the Shoshone River that represent some parts of the several Corbett masses we describe here. In the caption to their Figure 7, Nelson *et al.* (1972) suggested that these masses are small "fault blocks" that have slumped from former positions.

We know of only three ways by which rocks/debris of a considerable thickness lying at one level might be lowered intact (*i.e.*, "let down") to rest atop rocks at a lower level; in this instance up to 400 m lower. In all but one case (normal faulting), the result would be to greatly dilute the volume of displaced Paleozoic carbonate debris relative to that of the material of the surface area covered and/or rock debris of the underlying Willwood Formation that would have necessarily accompanied the Paleozoic rocks owing to the nature of the transport.

(1) **Creep** of the debris down a relatively planar inclined surface, such as may be seen with the Paleozoic carbonate rock debris occupying the tops of the high pediment making up the large area south and southeast of McCullough Peaks (**Figure 22**). The effect of surface creep is the clear dilution of the volume of carbonate rock relative to that at its source by the spreading of a limited volume of rock debris over what becomes an increasingly large surface area the farther the rock is moved from source.

(2) **Mass-wasting**, including *slumping, earth-flowage*, and/or *landsliding*. *Slumping* would require rock failure of the Willwood bedrock underlying intact debris at the top of the Peaks through the backward rotation of slump blocks—a phenomenon not seen in any of the Paleozoic debris. *Earthflowage* implies the flow of a wet debris matrix, for which there is also no field evidence in the Corbett masses. Earthflowage is evident in the Heart Mountain/McCullough Peaks area; however, its aspect (*e.g.*, two earthflows in the vicinity of 44° 41' 43" N, 109° 05' 10" W at Heart Mountain—**Figure 41**; one of several earthflows at the top of McCullough Peaks, at 44° 34' 25" N, 108° 50' 10" W, and another on the north side of the Peaks, at 44° 35' 26" N, 108° 49' 30" W (**Figure 42**) is quite different from that seen in the Corbett masses. *Landsliding* would require the collapse of a mass of intact Paleozoic carbonate debris from a source near the top of the Peaks (underlying Willwood rocks included in the landslide or not), fall of the debris for some distance, and the ultimate comminution of the debris and its spread over a large area beneath the area of collapse. Landsliding from the top of McCullough Peaks is also an unsatisfactory solution for the origin of the Corbett masses because the debris that occurs at a level hundreds of meters below its presumed source is about equally as thick as the Paleozoic debris atop the Peaks, and because no Willwood rock is mixed with the comminuted Paleozoic carbonate rock debris.

(3) **Normal faulting** would require massive displacement of the underlying Willwood Formation, for which there is no field evidence.

Source of Sturzstrom Debris

Geomorphological considerations indicate that the Heart Mountain/McCullough Peaks Sturzstrom event took place during a phase of intense regional erosion at some time after the Pliocene/Pleistocene boundary (boundary at ~2.58 Ma; Gibbard and Head, 2009), see below. An enormous volume of HMMPS landslide debris has been removed by erosion, including most of the debris between Heart Mountain proper and McCullough Peaks, and all of that west of Heart Mountain, in the only logical direction leading to the sturzstrom source area. A direct bearing passing through the middle of the landslide debris on McCullough Peaks westward to the central block of Madison limestone on Heart Mountain is about 293°, and continuance of that bearing passes through the overlook at Dead Indian Summit on the Chief Joseph Scenic Highway (elevation of 8060 feet = 2457 m; see Heasler *et al.*, 1996).

The Natural Corral area (**Figure 43**) north of Pat O'Hara Mountain (Pierce, 1965b; Pierce and Nelson, 1968) is is a steeply east-sloping stripped surface developed on the Permian Park City and Triassic Chugwater formations. This surface is bounded on the west and south in part by allochthonous masses of Paleozoic rocks of varying size of the late early Eocene SSADF. Some of the masses lie on the "ramp" between Pierce's "surface of tectonic denudation" and his former Eocene land surface, whereas others lie on the former land surface itself, having passed over the ramp. We hypothesize that the source of the HMMPS deposit was in this area and that the landslide debris originated in allochthonous SSADF blocks that occupied this former Eocene land surface in the area north of Pat O'Hara Mountain and, very likely, were distributed across other parts of the land surface to the south, in the vicinities of Bear Springs and Chalk Mountain, at the head of Rattlesnake Creek, and on and south of Pat O'Hara Mountain (Pierce and Nelson, 1968).

About 20-30 meters of powdery and granular carbonate debris supporting a tightly-packed welded breccia of angular clasts (**Figure 30**), identical to that forming most of the Corbett masses and much of the landslide debris on McCullough Peaks, and a carbonate/silicate breccia, is exposed in the roadcut on the Dead Indian Summit overlook (**Figure 31**), and above it at 44° 44' 37" N, 109° 22' 54" W and environs. The surfaces below and east of the Dead Indian Summit overlook are strewn with Paleozoic carbonate debris, from a rubble of rock flour and fine carbonate sand and granules to brecciated, angular blocks of dolomite and micrite a few centimeters up to 0.5 m in diameter. It is unclear whether this breccia resulted from movement on the SSADF or the HMMPS; however, we postulate that its position at the eastern extremity of the ramp indicates it was produced by SSADF allochthons as they cleared the ramp incline. X-Ray Diffraction analysis of the breccia (Appendix I) indicates that movement was rapid, in contrast to having taken about a million years, as postulated by Hauge (1982), or >2 Ma, as suggested by Hiza (1999b) for the HMDF.

> *"The strong shear strain in the quartz from near-fault samples requires significant shear stress, and can only be generated with rapid movement. It is not possible to create the strong shear strain with gradual movement. Movement that takes place over millions of years would give weak shear strain that would quickly rebound over the same millions of years."* (Mark Mathison, 2023, written communication to T. Bown regarding XRD analysis by Dr. Dan Hummer)

The massive, allochthonous, so-called Dead Indian Block (Voight, 1974) actually consists of a mass of several juxtaposed blocks that lie variously on both the transgressive ramp and Pierce's former land surface. The easternmost portion of the southeastern extension of this mass lying atop the Triassic Chugwater Formation on the former land surface is downfaulted on the west with an upfaulted mass of Devonian Jefferson and Three Forks formations on its eastern face (**Figure 6**). This small block (at 44° 42' 21" N, 109° 21' 01" W) is possibly a remnant of the head of the landslide scarp of the HMMPS. Unfortunately, the prodigious amount of erosion that has taken place in the time between emplacement of HMMPS debris and the present day has obliterated most geologic evidence of the precise location of the origin of the sturzstrom deposit.

Magnitude of Displacement (Run-out/Extent), and Mechanism of Movement

The length of the displacement of the HMMPS debris (termed the "run-out" for exceptionally large landslides), originating in the Natural Corral area above Sunlight Basin and ending far east of the East Peak of McCullough Peaks (to about 44° 30' 57" N, 108° 40' 30" W) is approximately 40 miles (about 65 km, **Figure 44A**), about the same as the run-out of the Enos Creek-Owl Creek Debris-Avalanche (ECOCDA; Wilson, 1975a; Bown, 1982a, 1982b; Bown and Love, 1987—see below and **Figure 44B**), some 60 miles (96 km) to the south. The presence of boulders of Bighorn dolomite 7.2

miles (11.8 km) north of Heart Mountain (**Figures 34 and 44A**) attests to an immense lateral splay of sturzstrom debris accompanying the run-out.

The eastern margin of the major part of the McCullough Peaks pediment is highly dissected (**Figure 22**) and its easternmost reach (at approximately 44° 32' 55" N, 108° 45' 37" W, some 3.27 miles = 5.26 km southeast of East Peak) is mantled with a thin veneer of carbonate rock flour and carbonate granules, pebbles, and sand-size grains. Although some of this material may have been transported down the pediment slope by surface creep, the nature of the deposit suggests instead that a dense cloud of finer airborne carbonate debris was propelled down the pediment surface for a considerable distance beyond the main sturzstrom deposit.

Bridger Butte is a flat-topped hill with a surface area of about 13 hectares, situated approximately 11.6 miles (= about 18.7 km) southeast of the easternmost margin of the McCullough Peaks pediment. Geomorphological relationships suggested to us that Bridger Butte may be an isolated remnant of the McCullough Peaks pediment; however, the composition of gravels capping the butte (Table IV) demonstrates the feature is instead a remnant of a high terrace of Dry Creek. The easternmost distal preserved remnant of the Heart Mountain-McCullough Peaks Sturzstrom deposit covers an elongate pediment outlier (**Figures 45 and 46**) approximately 10.4 miles (17 km) southeast of Middle Peak of McCullough Peaks and 6.85 miles (11.2 km) northwest of Bridger Butte, at about 44° 30' 57" N, 108° 40' 30" W. This deposit consists of a veneer of up to 20 cm of angular limestone and dolomite clasts, and with one boulder seen that measures 32 cm in diameter (**Figure 35**).

The northernmost extent of the HMMPS deposit east of the Shoshone River lies at about 44° 37' 34" N, 108° 45' 12" W, on a surface marked as "Pediment with limestone debris derived from Heart Mountain fault block" by Pierce (1985). We believe this material represents the most distally preserved northeastern remnant of the original HMMP sturzstrom deposit and one that, like the most distal deposit north of Heart Mountain itself, must have reached its distal extremity through seismically assisted transport.

The dimensions of both Wyoming mass-movements (HMMPS and ECOCDA) dwarf those of the Baga Bogd landslide of Mongolia (20 km run-out; Philip and Ritz, 1999) and the Saidmarreh landslide of southwestern Iran (16.1 km run-out; Harrison and Falcon, 1938; Roberts and Evans, 2013), and the Heart Mountain/McCullough Peaks deposit records by far the largest recognized sturzstrom event on earth (**Figure 44B**). Landslide debris of the HMMPS descended from an elevation of about 8,150 feet (2,485 m) at its proximal part to about 5,230 feet (1,595 m—the estimated elevation of the bed of the Shoshone River at the time of landsliding), engulfed the Shoshone River valley, and swashed upwards about 1,320 feet (402 m) to the top of the McCullough Peaks eminence where it swarmed over a rugged badland topography and finally descended at least an additional 1,259 feet (384 m) before the leading edge of the landslide debris came to rest on a southeasterly sloping pediment surface.

The largest intact carbonate blocks on McCullough Peaks were probably rafted to their current locations atop a slurry of debris, as indicated by their positions above the finer detritus (largely breccia). The upward migration of coarser clasts—even enormous slabs such as the 17,150 m^3 triangular block near East Peak—and their "rafting" atop finer debris is a common and signal feature of immense landslides (*e.g.*, Cruden and Hungr, 1986; Hewitt, 1988; Fauque and Strecker, 1988; Davies *et al.*, 1999). Pollet and Schneider (2004) observed that blocks riding atop the (much smaller) Flims sturzstrom deposits reached up to 1 X 10^6 m^3 in volume. Davies *et al.* (1999:1102) concluded that large clasts in the surface layer of a landslide deposit indicate that disaggregation and comminution took place beneath the surface of the deposit during movement, and they observed that larger blocks in the subsurface of the deposit were commonly shattered but "undisaggregated." Such shattered, yet undisaggregated blocks of limestone and dolomite weather from the finer debris matrix at several places on McCullough Peaks.

Shreve (1968) recorded several examples of landslide debris climbing the slopes of stream valleys opposite source (see also: Heim, 1883, and Oberholzer, 1933, for the Flims, Switzerland slide [Figure 47]; Harrison and Falcon, 1938, for the Saidmarreh, Iran slide; Kent, 1965, for the Frank/Turtle Mountain, Canada slide; and Voight, 1978, for the Gros Ventre, USA slide). The largest of those, the Saidmarreh rock avalanche of southwest Iran (Harrison and Falcon, 1938; Shoaei and Ghayoumian, 1996), is most closely analogous to the HMMPS, with a reach of about 16.1 km (Roberts and Evans, 2013), and a height climbed up the opposite valley of about 1,965 feet (~ 600 m).

The relative thinness of the most distal debris atop McCullough Peaks as compared with its great horizontal reach is also a characteristic of the largest known landslips (*e.g.*, Kent, 1965). About 90% of the Paleozoic debris on McCullough Peaks is comminuted; *i.e.*, it has been reduced to an angular pebble to granular and powdery rubble. These characteristics demonstrate that the Heart Mountain/McCullough Peaks displacement event was a sturzstrom (Heim, 1882)—a landslide with a relatively great horizontal runout—commonly reaching 20X-30X the vertical displacement. Regarding those deposits, (Hsü, 1975:130) observed that a sturzstrom is:

> "... a stream of very rapidly moving debris derived from the disintegration of a fallen rock mass of very large size; the speed of a sturzstrom often exceeds 100 km/hr, and its volume is commonly greater than 1×10^6 m^3."

and Heim (1932, as interpreted by Hsü, 1978:92):

> "... recognized that large sturzstroms flow farther than expected because of a reduction in their internal friction, and that this reduction is velocity dependent."

Pollet and Schneider (2004) recorded that the movement of debris transported via sturzstrom evolves into a rapid granular flow by a process they termed "dynamic disintegration."

Horizontal displacement of the Heart Mountain/McCullough Peaks landslide debris was 44X vertical displacement if measured from the gouge near Dead Indian Hill to the base of debris above the Shoshone River, and about 72X vertical displacement if measured from the Dead Indian Hill area to the top of the end-of-runout pediment outlier at 44° 30' 57" N, 108° 40' 30" W. We contend that the vast runout of the HMMPS was assisted by sustained, earth-shaking seismic activity. Preliminary rock analyses indicate HMMPS movement was both catastrophic and rapid (Mathison *et al.*, 2025).

ꙮ Age of Displacement
Geomorphological Considerations

Previous to this study it was assumed that the allochthonous masses of Paleozoic carbonate rocks forming Heart Mountain and capping McCullough Peaks were emplaced as part of the late early Eocene Heart Mountain Detachment Fault (HMDF=our SSADF). The key to the age of emplacement of these masses, however, lies in determining the ages of the surfaces upon which the masses lie. On Heart Mountain, the immense, more-or-less intact central block of Mississippian Madison Limestone (about 7.3×10^6 m^3) is tilted about 4°-6° to the north, a position that resulted from the block having entered an erosional declivity, rather than continuing to move eastward across a gently sloping Eocene land surface confluent between the Natural Corral area and the top of McCullough Peaks. The inclined surface at the base of the easternmost debris on Heart Mountain indicates that the hypothetical extension of this declivity farther to the east at the time of emplacement was yet deeper than any part of today's preserved part of that surface beneath the blocks; *i.e.*, it can be postulated that a declivity similar to that of the present-day Shoshone River valley existed between Heart Mountain and McCullough Peaks at the time of movement. Similarly, masses of allochthonous Paleozoic carbonate debris on the McCullough Peaks side of the Shoshone River lie on remnants of a surface that descends westward over 500 m into the Shoshone River Valley.

The high point of McCullough Peaks (Middle Peak of Hewett, 1920, at an elevation of 6546 feet = 1996 m) is a point on an erosional surface on the lower Eocene Willwood Formation and not on allochthonous material, demonstrating that a pronounced erosional topography existed at the top of the Peaks during emplacement. The youngest fossil vertebrates collected from the topographically highest Willwood exposures in the McCullough Peaks area were obtained about 80 meters beneath the top of Middle Peak. From collections of fossil mammals at the University of Michigan Museum of Paleontology, Dr. Will Clyde (personal communication of 01/30/2018) observed (brackets are ours):

> *"My guess is that the top of the Willwood Formation [exposed on] McCullough Peak [highest elevation on Willwood Formation in the McCullough Peaks area] is in the middle of [faunal horizon] Wa-6"*

(*e.g.*, Gingerich, 1980, 1991, for faunal zonation, modified by Chew, 2005; see also Clyde, 2001). Dr. Amy Chew (personal communication of 09/06/19) notes that the Wa-6 faunal horizon is coincident with about the 430-590 meter interval in the continuous 770 meter-thick Willwood Formation section in the central Bighorn Basin (measured by Bown, in Bown *et al.*, 1994), and the stratigraphically youngest Willwood rocks in the McCullough Peaks area occur not at or near the topographically highest part of McCullough Peaks (atop Middle Peak), but in the southeastern McCullough Peaks composite section (of Clyde, 2001), at a much lower elevation and several miles to the southeast. These data demonstrate that the allochthonous masses on the Peaks were strewn across Willwood (lower Eocene) rocks that had been folded and deeply eroded prior to the HMMPS displacement.

The structural folding of Willwood rocks in the McCullough Peaks area resulted from the rapid structural elevation of the Cody Arch (of Sundell, 1990, 1993). According to Lillegraven (2009), that orogenic movement immediately preceded displacement of Paleozoic rocks by the Heart Mountain Detachment Fault (our SSADF); *i.e.*, it occurred immediately prior to 49.5 Ma in the late early Eocene. Lillegraven (2009:76) notes that:

> *"... the amount of syntectonic erosion of strata along the Bighorn Basin's western margin associated with uplift of the Cody Arch must have been prodigious."*

The question remains whether or not that prodigious late early Eocene erosion resulted in the topographic surface upon which lies the allochthonous Paleozoic debris capping Heart Mountain and McCullough Peaks.

If one is to accept the premise that the Paleozoic carbonate allochthons atop McCullough Peaks were emplaced during the late early Eocene, it is necessary to discard a great deal of circumstantial evidence that they were not. Based on sediment accumulation rates of 0.90 – 1.58 feet/1,000 years (= 0.27 – 0.48 m/1,000 years) calculated for the Willwood Formation (Westerhold et al., 2018), in the 49-odd million years between the time of SSADF movement and the present day, 44,100 – 77,420 feet (= 13,445 – 23,603 m) of sediment *could have been* deposited in the Bighorn Basin. Conversely, based on sediment erosional rates of one foot (about 0.305 m) per 1,500 years calculated for the time between the late Miocene and the present day (J.D. Love and H. Walsh, 1981, personal communication), in the 49-odd million years between the time of SSADF movement and the present day, a total of up to 32,667 feet (9,959 m) of sediment *could have been* eroded from the Bighorn Basin. Of course, neither of these scenarios took place; however, there is ample evidence that, following the early Eocene, the Bighorn Basin and several other Rocky Mountain intermontane basins continued to fill with sediment up to the level of the incised subsummit surfaces of their adjacent mountain ranges (*e.g.*, Mackin, 1937, 1947; Love, 1960; McKenna and Love, 1972; Ritter, 1967, 1975; McKenna, 1980; Giegengack *et al.*, 1986, 1988; Mears, 1993; J.R. Malone *et al.*, 2022), and that excavation of those high-level basin-fill deposits probably began in the late Miocene.

What a remarkable circumstance would obtain if, after whatever episodes of deposition and erosion took place in the northern Bighorn Basin following emplacement of comminuted HMDF carbonate debris on McCullough Peaks 49-odd million years ago, much of the allochthonous carbonate debris remained unindurated/uncemented following its burial by what was likely to have been a considerable thickness of middle Eocene and younger pyroclastic and volcaniclastic sediments. Instead, much of the allochthonous material on the Peaks is today a friable, largely unindurated rubble of mostly brecciated angular granule-to-cobble-size micritic and dolomitic clasts, swimming in a relatively unindurated carbonate flour and forming a matted, highly permeable veneer that lies on relatively impermeable clay-rich mudstones of the lower Eocene Willwood Formation. Those mudstones act as an aquitard/aquifuge such that numerous verdant zones of vegetation (**Figure 48**) and several springs (*e.g.*, Markham Spring, at 44° 34' 15" N, 108° 49' 32" W, described by Hewett [1920], and several others on the McCullough Peaks rise) have their origins with water concentrated at the contact of the permeable allochthonous carbonate breccia with the underlying Willwood mudstones.

On both Heart Mountain and McCullough Peaks, the HMMPS deposit lies upon a thinly mantled bedrock pediment (*e.g.*, Twidale, 1981; Bourne and Twidale, 1998; Applegarth, 2004). Beneath both physiographic features, the bedrock platform consists of the lower Eocene Willwood Formation, and in both instances the pre-sturzstrom pediment surface of the bedrock platform exhibits a veneer of a few centimeters of clastic debris. At Heart Mountain, this material is mostly chert and quartzite pebbles (Table IV), and on McCullough Peaks the debris is rounded granules and pebbles of volcanic detritus mixed with sandy and silty loess. Lateral stream corrasion (lateral planation of Johnson, 1932) and sheetwash (*e.g.*, McGee, 1897; King, 1949) were probably effective in development of the Heart Mountain and McCullough Peaks pediments due to the easily erodible character of the majority volume of the bedrock sediment (Willwood mudstone).

If the allochthonous masses on McCullough Peaks were emplaced in the late early Eocene, equally remarkable would be the fact that erosion since the Miocene has so precisely left the top of the surface veneer of the debris on the McCullough Peaks pediment parallel with the underlying pediment surface. Moreover, isolated high pediment surface remnants that are inclined away from the base of the allochthonous debris on Heart Mountain yet, by extension, were confluent with it, have been extensively dissected by streams and preserve upper surfaces capped by allochthonous debris. In other words, all of the pediment surfaces sporting landslide debris on Heart Mountain and McCullough Peaks conform in every way with pediment surfaces in the surrounding region that exhibit no allochthonous debris; *i.e.*, all the surfaces fit most naturally into the geomorphic landscape produced during the current

Pleistocene and Recent erosional cycle. Ruler Bench (**Figure 49**) is a high, dissected pediment off the Beartooth Mountains southeast of Red Lodge, MT that is analogous to the elevated surface that we conceive was engulfed farther south by HMMPS debris in the early Pleistocene.

Lastly, no exceptionally thick, mature paleosol is developed on the McCullough Peaks pediment, nor was the same formed atop any of the other major high-level pediment surfaces in the region. This circumstance negates the hypothesis that the current landscape is a relict of the late early Eocene that has survived in a state of stasis (*i.e.*, no appreciable deposition upon it, nor erosion of it) since HMDF detachment faulting.

Erosional Rates

Datable tuffs interbedded with erosionally elevated stream terrace deposits in the eastern Bighorn Basin have been used to estimate erosion rates of the Bighorn Basin sedimentary fill (Palmquist, 1978); these being approximately 0.2 meters/1,000 years for the interval 2.0-0.6 Ma, and 0.16 meters/1,000 years for the interval between 0.6 Ma and the present. These rates are closely coincident with the 1 foot (= 0.305 m)/1,500 years calculated by Love and Walsh (1981) for the 600,000 to 6,000,000-year interval in the southerly adjacent Wind River Basin. Using the Palmquist determinations, excavation of the top of the largest of the Corbett masses, which lies 425 m above the level of the adjacent Shoshone River, is estimated to have begun at approximately 2.245 Ma. In contrast, the Love and Walsh (1981) formula suggests excavation began at about 2.091 Ma. Table III lists the estimated ages of other relevant erosional surfaces in the vicinity of the Heart Mountain/McCullough Peaks Sturzstrom Deposit and the Enos Creek/Owl Creek Debris-Avalanche (ECOCDA).

Terraces Along Dry Creek

Seven major streams traverse most or all of the Bighorn Basin from west to east and merge with the north-flowing Bighorn River on the eastern side of the basin (**Figures 1 and 50**). From north to south, these are: the Shoshone River, Dry Creek (*e.g.*, Robinove and Langford, 1963) = Oregon Coulee, in part, of the older literature (*e.g.* Hewett, 1926); The Greybull River (= Gray Bull River of the older literature; *e.g.*, Sinclair and Granger, 1912); Fifteenmile Creek (= Dry Cottonwood Creek of the older literature; *e.g.*, Mackin, 1947, Fig. 5); Gooseberry Creek, Cottonwood Creek, and Owl Creek. All but Dry Creek and Fifteenmile Creek are perennial streams and find their headwaters in the volcanic Absaroka Range. Being intermittent streams, both Fifteenmile Creek and Dry Creek are generally dry or unflowing, have no headwater access to mountain snowfields and, more importantly, have no primary water/sediment sources in outcrops of *in situ* Paleozoic or volcanic rocks. At least the lower reaches of Dry Creek and the Greybull River owe their linearity to control by basement structure (Kraus, 1992; Bown *et al.*, 2016); Dry Creek probably to the basinward extension of the Shell Creek Lineament, and the Greybull River to a propagation of the Shell Creek lineament or possibly to the Florence Pass Lineament (Hoppin and Jennings, 1971; Hoppin, 1974; Kraus, 1992; Bown *et al.*, 2016).

The headwaters of Fifteenmile Creek lie in badlands of the mostly alluvial Fort Union (Paleocene) and Lance (uppermost Cretaceous) formations in the Hole-In-The-Ground *cul-de-sac* southeast of the town of Meeteetse, WY. The North Fork of Dry Creek has its source in the Upper Cretaceous Cody and Mesaverde formations, sedimentary rocks of marine, beach, and/or nearshore continental origin outcropping just west of the Oregon Basin (at about 44° 19' 32.83" N, 109° 00' 18.98" W; see, *e.g.*, Hewett, 1926); whereas the South Fork of Dry Creek heads at a spot on the Meeteetse Rim, about 45 m above the elevation of the bed of the closely adjacent Meeteetse Creek, south of the Meeteetse Rim, in Upper Cretaceous Lance or Paleocene Fort Union rocks several miles west-southwest of the Oregon Basin (at about 44° 16' 20.54" N, 109° 03' 56.95" W). The Meeteetse Rim is a prominent, more-or-less west-to-east aligned rise that separates the drainage basin of Sage Creek and the North and South Forks of Dry Creek (on the north) from the drainage basin of Meeteetse Creek, Spring Creek, Rawhide Creek, and the Greybull and Wood Rivers (on the south) (**Figure 50**).

Today's headwater area of the North Fork of Dry Creek has no access to sediment clasts derived from volcanic rocks or from Paleozoic carbonate rocks of *in situ* (*i.e.*, non-allochthonous) origin, and the headwater area of the South Fork of Dry Creek has no access to volcanic rocks or Paleozoic carbonate rocks of either *in situ* or allochthonous (SSADF or HMMPS) origin. The conundrum posed by this circumstance is that the lowermost (youngest) terrace deposits of Dry Creek east of the confluence of the North and South Forks, as well as most terraces of the North Fork and the South Fork of Dry Creek west of that confluence within the Oregon Basin (Table III) contain abundant volcanic clasts and a significant percentage of carbonate clasts derived from Paleozoic sources (Table IV).

After breaching the Absaroka Range, today's Greybull River flows to the east up to a point a few miles southwest of the town of Meeteetse where its course bends to the northeast. About three miles southwest of the town of Burlington, the northeasterly course of the Greybull shifts abruptly to the east-southeast at a point Mackin (1936, Figure 1) termed point "D". Mackin proposed that the lower drainage of the Greybull River once coincided with that of today's lower drainage of Dry Creek and that the ancient Greybull River joined the Big Horn River where Dry Creek joins the Big Horn River today. Headward erosion to the west by a smaller stream (Mackin, 1947, Fig. 2) captured the Greybull River at his (1936 Fig. 1) point "D" (see point marked "PC" on **Figure 50**).

If Mackin's (1936) construct of the capture of the Greybull River is accurate, there should be no terrace deposits containing volcanic clasts along the drainage of Dry Creek west of its point of post-capture abandonment because Dry Creek's source of water originating in the Absaroka Range was cut off. Yet, low level (relatively young) terrace deposits along Dry Creek west of a north-south locus drawn through Mackin's point "D" are dominated by clasts of volcanic rocks and contain an appreciable fraction of clasts of Paleozoic carbonate rocks.

Merrill (1974) recognized only two terrace levels along Dry Creek west of the confluence of the North and South Forks, and three levels on Dry Creek between that juncture and its joining with the Bighorn River. We identified at least five levels of "cut-in-rock" terraces (terminology of Merrill (1974; his Figure 4) along Dry Creek in the Oregon Basin (**Figures 51 and 52**), with each of the levels separated by bedrock risers. Each of these terrace deposits is dominated by volcanic clasts and contains 1-15% of clasts of Paleozoic carbonates, with one sample from the Elk Butte terrace deposit (the highest terrace deposit in the Dry Creek drainage) containing about 35% Paleozoic carbonates (Table IV). Where do these clasts come from?

These observations support our hypothesis that debris of the HMMPS deposit dammed the Shoshone River, probably both west and east of the Cody Arch (Rattlesnake/Cedar Mountain, see below). Lakes formed behind the dams and the lake waters rose, flooded the valley of Sage Creek, excavated the low swale that contains today's Road 3 FK (between the Corbett area and the North Fork of Dry Creek), and dumped water into what became the North Fork of Dry Creek in the northern part of the Oregon Basin.

Keeping in mind that the elevations of the beds and floodplains of all of these streams were several hundred feet higher when many stream captures and the HMMPS took place, it is also probable that dammed floodwaters entered the Oregon Basin farther to the south where the highest (and oldest) terraces containing volcanic and Paleozoic carbonate clasts occur (*e.g.,* on Elk Butte, **Figures 50** and **52**; Table IV). Several dry, elevated swales are developed in the areas of Sulphur Creek and Spring Creek, and the position of the long, west-east swale containing Monster Lake and Wiley Lake indicates that Sage Creek, which turns sharply north at its juncture with this swale, was once confluent with the North Fork of Dry Creek and that Sage Creek was captured by the headward erosion of a north-flowing drainage. A post-HMMPS landslide deposit (*e.g.,* at 44° 21' 57" N, 109° 09' 14" W) has blocked another swale that once may have joined dammed post-HMMPS lake waters with Sage Creek at about 44° 20' 56" N, 109° 06' 21" W.

A detailed study of the erosional history of even this restricted area of the northwestern Bighorn Basin, much as it might lead to a clearer history of the HMMPS, is a several years' field study that lies well beyond the scope of this project (see Palmquist, 1983) but, as Mackin (1937:837) cogently observed:

> *"Determination of the identity of the streams that formed the abandoned valleys and the high, gravel-capped interstream benches of the Bighorn Basin is essential to the interpretation of these features."*

However, the data at hand are sufficient to offer some intriguing suggestions regarding the relative ages of two episodes of stream capture and the Heart Mountain/McCullough Peaks Sturzstrom:

Having no carbonate clasts, the Fenton Pass Formation—the terrace deposit atop Tatman Mountain—pre-dates the capture of the Wood River by the Greybull River. The head of the Wood River accesses clasts of Paleozoic carbonates derived from roof-pendant masses forming the top of Dollar Mountain (see below) and, prior to its capture by the Greybull River, the Wood River probably flowed more-or-less directly east into and across the central Bighorn Basin, as evidenced by volcanic cobbles occurring on several surfaces in the tributary drainage of the *cul de sac*-bounded head of Fifteenmile Creek (*e.g.,* on a terrace remnant at 44° 09' 14" N, 108° 14' 47" W).

The age relationship of the Fenton Pass Formation (= the Tatman Mountain Surface) to the capture of the Greybull River is uncertain. Lying slightly west of Mackin's (1936) Point "D", the Fenton Pass volcanic gravels could either pre-date or post-date the capture of the Greybull River. Red Butte, however, is a volcanic gravel-capped peak at an elevation of 5,151' (1,570 m) that lies about 22 km southeast of the east end of Tatman Mountain and considerably east of Mackin's Point "D". Volcanic gravel atop Red Butte indicates that deposit post-dates the capture of the Greybull River. The Red Butte gravel has an estimated age of between 1,372,131 and 1,515,000 years (Table III), so the capture of the Greybull River occurred at some time prior (and probably well prior) to 1,372,131 years.

The presence of Paleozoic carbonate clasts in certain terraces along only the upper Dry Creek drainage indicates that those terraces formed after the capture of the Greybull River. Because the availability of those carbonate clasts was almost certainly due to the HMMPS deposit, the capture of the Greybull River preceded the HMMP Sturzstrom (pre-2.08 Ma).

These data suggest that a major Bighorn Basin erosional episode was triggered by the same event that instigated the HMMP sturzstrom, and that a second erosive episode of unknown origin began between 1.5 and 1.3 Ma (erosion from the levels of the Elk Butte and Red Butte surfaces, possibly = Mesa Falls Tuff eruption at 1.3 Ma, *e.g.,* Izett and Wilcox, 1982).

❧ Triggering Mechanism

Keefer (1984) concluded that most landslides, including those occurring extraneous to volcanic settings, are probably instigated by earthquakes (see also Terzaghi, 1950; Solonenko, 1977; Ui, 1985; Crozier, 1992; Jibson, 1996). Whereas an immense volume of Paleozoic carbonate rocks was displaced by an exceptionally violent megathrust earthquake resulting in the late early Eocene Shoshone/Sunlight/Abiathar Detachment fault—formerly termed the Heart Mountain Detachment Fault—we offer field evidence indicating a much younger emplacement age for all the allochthonous rocks atop Heart Mountain, McCullough Peaks, and intervening areas, as was advocated by Bown and Love (1989).

Geomorphological/radiometric determinations of the estimated rates of excavation of the sedimentary fill of the Bighorn Basin suggests an age of 2.091-2.245 Ma for emplacement of the Heart Mountain/McCullough Peaks Sturzstrom deposit, an age approximately contemporaneous with the Enos Creek-Owl Creek Debris-avalanche—another probably earthquake-induced displacement that Bown and Love (1987) concluded resulted from the collapse of the Island Park Caldera and was coincident with deposition of the ~2.08 Ma Huckleberry Ridge Tuff (Christiansen and Blank, 1972; Christiansen, 2001; Swallow *et al.*, 2019). Field evidence indicates that exceptionally violent seismicity associated with one of the several silicic Huckleberry Ridge Tuff eruptions also induced movement of SSADF-emplaced Paleozoic blocks once present on Pierce's (*e.g.*, 1960) former land surface between Pat O'Hara Mountain and Dead Indian Hill, causing them to collapse eastward into the Shoshone River declivity and swash up the rise on the east side of the river to cover rocks of the eroded lower Eocene Willwood Formation and the high pediment surface below the easternmost crest of the McCullough Peaks eminence. The occurrence of boulders of Paleozoic carbonate rock several miles distant to the main locus of the displacement suggests that the immense runout and splay of the HMMPS deposit was seismically assisted.

The Island Park caldera is immense and extends 60 miles (~97 km) from SW to NE and 40 miles (~64 km) from SE to NW (Fritz and Thomas, 2011; Hendrix, 2011). The volume of ash ejected in the Island Park caldera eruption is estimated at 600 cubic miles (2,483 km^3), and this was the first of three such violent eruptions under or near what is now Yellowstone National Park. The second occurred at 1.3 Ma and resulted in the Mesa Falls Tuff and the much smaller Henrys Fork caldera which lies within the southwest part of the Island Park caldera. The third, and most recent massive violent eruption occurred at 0.64 Ma, yielded the Lava Creek Tuff and produced the Yellowstone caldera which overlaps the northeastern half of the Island Park caldera and extends farther northeast. The Yellowstone caldera marks the present day position of what is known as the Yellowstone hotspot track—a trail of extinct volcanic centers or calderas that extends southwestward to near the junction of the borders of Idaho, Nevada, and Oregon.

To get a sense of the magnitude of the Island Park Caldera it is illustrative to compare it with some of the physical characteristics and aftermaths of two past super volcanic eruptions: 1) Mount Tambora, which exploded violently in April of 1815 is the largest eruption in recorded human history (Greshko, 2016); and 2) the eruption of the Toba volcano approximately 73-75 Ka, the largest known eruption in the last 25 Ma (Yale Leitner Observatory). Both volcanoes are part of the Sunda Arc, the island arc chain of volcanoes that forms a portion of the Indonesian archipelago. Mount Tambora is located to the southeast, in the Lesser Sunda Islands on Sumbawa Island (Smithsonian "Tambora: Program"; Oppenheimer, 2003), and Mount Toba is located approximately 2,400 km (1,490 miles) to the northwest, on the northern part of the island of Sumatra (Oppenheimer, 2002). The tectonic setting is the same for both volcanoes as they lie roughly 325 km (200 miles) inboard of the Java Trench (Sunda megathrust), a north to east-dipping subduction zone (Foden and Varne, 1980). At Mount Tambora the India-Australian Plate is being subducted beneath the Sunda microplate, whereas at Mount Toba the India-Australian Plate is being subducted beneath the Burma microplate extension of the greater Eurasian Plate (Chesner and Rose, 1991).

Although Tambora is the largest volcanic eruption in recorded human history, its caldera dimensions (6 x 7 km = 3.7 – 4.3 miles) are tiny compared to those of Island Park (approximately 100 x 66 km = 60 x 40 miles). The gigantic Toba caldera (at 101 x 36 km = 62 x 22 miles) is much more comparable in size to the Island Park Caldera. Total ejecta for Tambora is estimated at 64 – 97 km^3 = 24-36 miles^3 (Oppenheimer, 2003), whereas that from Toba was approximately 2,800 – 3,800 km^3 = 670 – 900 miles^3 (Morris, 2013). These enormous volumes are comparable to or greater than the 1,640 km^3 = 600 miles^3 of ejecta estimated for the Island Park eruption. Toba's ash has been found up to 7,000 km (~4,350 miles) to the west, in East Africa (Lane et al., 2013). The eruption of Mount Tambora lowered its maximum elevation by 4,746 feet = 1,447 m (Encyclopedia Britannica, 2023).

In addition to the volume of ash disgorged by Mount Tambora, 10-100 million tons of sulfur was also released (Oppenheimer, 2003), creating a global stratospheric veil that precipitated extreme weather conditions in parts of the Northern Hemisphere. Anomalously cold temperatures were felt in the northeastern United States, maritime Canada, and Europe during 1816; the so-called "year without a summer". Average global temperatures decreased about 0.4° to 0.7° C (0.7° – 1.3° F), causing widespread crop failure in North America and Europe (Stothers, 1984). Plunging temperatures broke the monsoon cycle in India, triggering famines and a severe cholera epidemic, and cold temperatures and intense rain storms devastated rice paddies, causing starvation in China (Greshko, 2016).

The dramatic aftermath of the Toba eruption was the basis for the Toba catastrophe theory which maintains that the eruption ushered in a "volcanic winter" that lasted for several years, with a global decrease in temperature of between 3° and 5° C (5.4° - 9.0° F), and up to 15° C (27° F) in higher latitudes (Rampino et al., 2000). These harsh conditions were theorized to have contributed to a 1,000-year cooling episode that led to a genetic bottleneck in human evolution in central East Africa and India, in which the human ancestral population was reduced to an estimated 1,000 to 10,000 breeding pairs worldwide (Ambrose, 1998; Rampino et al., 2000). More recent studies, however, argue that the human population in East Africa and India was little affected by the eruption (Lane et al., 2013; Yost et al., 2018; Max Planck Institute, 2020).

The Yellowstone Caldera marks the present-day position of the Yellowstone hot spot track, a trail of extinct volcanic centers or calderas that extend southwestward to near the junction of the borders of Idaho, Nevada, and Oregon. The origin of the Yellowstone hotspot is controversial and a detailed analysis of the various theories of its origin lies beyond the scope of this report. However, because the volcanism attributed to the HMMPS displacement was much more explosive than the stages of volcanism associated with the SSADF, a brief summary is appropriate.

The Yellowstone hotspot apparently first appeared at 16.5 Ma (but see Camp and Wells, 2021) more than 400 miles southwest of the Yellowstone caldera with an eruption that created the McDermitt caldera, the first of more than 140 eruptions in what is now Idaho's Snake River Plain. The Columbia River flood basalts were also developed at this time and have been considered to be evidence of early volcanism related to the hotspot. Five other major volcanic super-eruptions resulted in calderas that became successively younger to the northeast: 1) Owyhee-Humbolt (13.8 Ma); 2) Bruneau-Jarbridge (12.5 Ma); 3) Twin Falls (10.8 Ma); 4) Picabo (10.3 Ma); and 5) the Heise caldera (6.65-4.45 Ma) that includes several small calderas. Subsequently, these calderas were filled by additional volcanics resulting in the relatively flat Snake River Plain (Fritz and Thomas, 2011).

The early, conventional view of the origin of the Yellowstone hotspot was that it formed over a relatively stationary plume from the deep core-mantle boundary similar to the formation of the Hawaiian Islands. The North American plate has moved southwestward over this plume in response to the continued opening of the Atlantic Ocean since the breakup of Pangea. More recently, this model has been challenged by proponents of an upper mantle origin. Christiansen *et al.* (2002) observed that seismic imaging and helium isotope ratios did not provide evidence for a deep mantle origin, and Zhou *et al.* (2018) suggested that volcanism associated with the hotspot is caused by the partial melting and eastward driven upper mantle flow of either the subducted Farallon plate and/or the more recent subduction of the Juan de Fuca plate—a separated remnant of the Farallon plate.

Other workers have attributed the origin of the hot spot track to either massive rifts or a tear in the Farallon slab. Glen and Ponce (2002) proposed that during the middle Miocene a series of rifts and other geologic features extending hundreds of miles across Nevada converge at a point near the Idaho/Oregon border that may be a point source of stress at the base of the crust related to the formation of the Yellowstone hotspot. Another model (Hendrix, 2011) suggests that a tectonic reorganization took place at about 20 Ma when a spreading center subducted beneath the North American plate resulted in a massive fracture that was propagated northeastward toward Yellowstone. The hotspot track was then formed as the thermal plume followed the leading edge of the propagating fracture. Liu and Stegman (2012) attribute Miocene and younger Yellowstone volcanism to a 17 Ma episode of Farallon slab tearing under eastern Oregon that created a slab gap at shallow depth that quickly ruptured to the north and south.

Hooper *et al.* (2007), Nelson and Grand (2018), Steinberger *et al.* (2019) and Camp and Wells (2021) renewed arguments for a mantle plume origin for the Yellowstone hotspot. Camp and Wells (2021) suggested that the hotspot is a "whole-mantle plume" at least as old as 56 Ma and was responsible for Paleocene and Eocene volcanism that created an oceanic plateau that was accreted to the North American Plate from 51-49 Ma. Nelson and Grand (2018) and Steinberger *et al.* (2019) presented a seismic tomography model that reveals an anomaly they interpreted as a whole mantle plume extending from the core-mantle boundary through the lower mantle to the Yellowstone hotspot at the surface, supporting Hooper *et al.* (2007) in ascribing the hotspot trend to plate motion over a deep-seated mantle plume.

Debate over hotspot origins notwithstanding, the eruptions that produced calderas such as those spewing the Huckleberry Ridge tuffs at about 2.08 Ma were exceptionally violent, much more so than volcanism attributed to events associated with the SSADF, and more explosive than any others in Absaroka Range geologic history, and even more so than any in recorded history with the possible exception of Mount Toba. The extreme violence of the eruptions resulted from the development of a highly viscous siliceous magma when the gabbroic plume rising through the aesthenosphere reached and began to melt the granitic continental crust. This viscous granitic magma contained abundant dissolved gases and exploded violently as rhyolitic lava flows. Volcanic eruptions associated with hotspots beneath oceanic crust (*e.g.*, the Hawaiian Islands) are much quieter because basaltic magma is much less siliceous and thus has a lower viscosity.

❧ Analogous Structures

The Heart Mountain/McCullough Peaks Sturzstrom deposit consists of: (1) massive (up to 7.3×10^6 m^3 in volume) blocks and other masses of intact, stratified carbonate rocks on Heart Mountain, nearest to the source of the displaced masses; (2) lesser blocks representing parts of individual beds (up to 17.15×10^3 m^3 in volume) atop masses of brecciated Paleozoic carbonate debris in the middle reach of the deposit (on McCullough Peaks, and in the Corbett masses); and (3) largely comminuted, granular debris supporting a few large carbonate blocks most distal to the source of the deposit (distributed across the McCullough Peaks pediment surface).

Several mechanisms have been proposed for the emplacement of sturzstrom deposits (*e.g.*, Shaller and Shaller, 1996). Detailed study of the internal structure of the HMMPS deposit is beyond the scope of this study; however, examination of its spatial distributional attributes indicates that the composition of the deposit—from massive blocks to granular particles—probably resulted from a shearing, slab-on-slab reduction of the mass during transport (Schneider *et al.*, 1999; Wassmer *et al.*, 2002; Pollet *et al.*, 2005), and the production of fine particles was due to the process of dynamic disintegration of the rock mass during movement (Pollet and Schneider, 2004). Complete disintegration of rock masses as large as the blocks on Heart Mountain would have resulted from violent collision of the masses with the west-facing cutbank rise of the Pleistocene Shoshone River, and the forward propulsion of ruptured blocks that were transported up the west side of McCullough Peaks further comminuted the debris by block-to-block, slab-to-slab, and grain-to-grain contact during movement. The resulting disintegrated granular material was propelled as a rock flow with some of the larger, surviving blocks being transported over the crest of the Peaks atop the mobile mass. This transport scenario mirrors that of the immense, yet considerably smaller Flims (Switzerland) sturzstrom (**Figure 47**), and is analogous also with the internal structure of the Saidmarreh Landslide deposit in the Zagros Range of southwestern Iran. The HMMPS deposit also shares some similarities with the immense volume of debris-avalanche detritus associated with the Mount Shasta volcano (Crandell *et al.*, 1984).

PERIPHERAL PHENOMENA
(T.M. Bown and A.J. Warner)

Squaw Buttes Allochthons

Squaw Buttes (**Figures 1 and 53**) are a pair of conical buttes perched atop the East Ridge Divide in the southwest-central Bighorn Basin. The lower parts of the buttes are composed of *in situ* shales and sandstones of the predominantly lacustrine lower and middle Eocene Tatman Formation (Sinclair and Granger, 1912; Rohrer and Smith, 1969; Bown *et al.*, 1994; Malone *et al.*, 2014), and the upper ~30% of the west butte and upper ~10% of the east butte consist of steeply inclined to chaotically disaggregated volcaniclastic sandstone, breccia, and conglomerate that Rohrer (1966) was the first to identify as allochthonous. Wilson (1970, 1975a) incorporated the Squaw Buttes allochthons in his Enos Creek Detachment Fault, and Bown (1982a, 1982b) included them in his expanded Enos Creek/Owl Creek Detachment Fault—later termed the Enos Creek/Owl Creek debris avalanche (Bown and Love, 1987). More recently, Malone *et al.* (2014) suggested that displaced volcanic rocks capping Squaw Buttes (their "Squaw Peaks") are the farthest-flung remnants of the Heart Mountain Detachment Fault (our SSADF), thus vastly increasing both the presumed maximum distance of HMDF allochthon transport (by about 50 miles = 80 km—Squaw Buttes lie about 50 miles southeast of Heart Mountain) and the total area affected by HMDF (from about 3,600 km^2 to at least 5,000 km^2). We present evidence that allochthonous rock capping Squaw Buttes is unrelated to movement associated with the HMDF (Shoshone/Sunlight/Abiathar Detachment Fault).

Malone *et al.* (2014:334-335) obtained a date of 50.53 + 1.59/-0.78 Ma from zircons in a clast taken from a "polymict volcanic cobble conglomerate" forming part of the allochthonous volcanic rocks at the top of West Squaw Butte. Those authors used this age to support their hypothesis that the Squaw Buttes allochthons were derived from the Cathedral Cliffs Formation or the Lamar River Formation ("early acid breccia" of Hague, 1899) of the northern part of the Absaroka Range, stating (2014:329) that:

> *"As no vent facies rocks of this age occur in the southern Absaroka Range, the 50.53 Ma age of the allochthonous volcanic rocks also indicates that these rocks were not likely emplaced as part of the late Cenozoic Enos Creek-Owl Creek debris-avalanche deposit."*

Overlooked is the likelihood that the vent-facies zircons (which occur in volcaniclastic rocks) were transported southward and deposited with other Aycross sediments by eolian or fluvial action concomitant with volcanic activity.

Volcaniclastic sediments of the Aycross Formation were derived largely from andesitic and trachytic rocks, and andesites and andesitic volcanic and volcaniclastic rocks are dominant in the Aycross Formation in its type area and elsewhere (Love, 1939; Bown, 1982a), with Aycross conglomerates containing abundant andesitic clasts. Whatever the origin of those clasts, it might well be the same as that of the zircon-dated volcaniclastics on Squaw Buttes.

The Malone *et al.* (2014) date of 50.53 + 1.59/-0.78 Ma for the allochthonous rocks on Squaw Buttes accords remarkably well with a date of 50.4 +/- 0.5 Ma determined by Love *et al.* (1976, and *in* Bown, 1982a) for the type section of the also fluvial volcaniclastic Aycross Formation in the northwestern Wind River Basin, and with a zircon date of 50.05 +/- 0.65 Ma from Aycross rocks near Togwotee Pass (Malone, Craddock, Garber, and Trela (2017). The latter two Aycross sections lie much farther south, far more distant from any Cathedral Cliffs or Lamar River outcrops, at the extreme southern edge of the Absaroka Volcanic Field, some 60 miles (~100 km) southwest of Squaw Buttes, and 25-30 miles (40-48 km) west-southwest of the Rhodes Allochthon of Bown and Love's (1987) Enos Creek-Owl Creek Debris-avalanche deposit. Therefore, the date of 50.53 + 1.5/- 0.78 Ma obtained by Malone *et al.* (2014:334-335) from the "polymict volcanic cobble conglomerate" at Squaw Buttes simply records the date of crystallization of the volcanic rock from which the dated clast was derived. The date in no way constitutes evidence that those rocks were emplaced on the East Ridge Divide as part of the Heart Mountain Detachment Fault (=SSADF).

Citing no supporting evidence, Sundell (1990), Malone (2000), and Malone and Sundell (2000) ascribed "early middle Eocene" (presumed Aycross age) and "middle Eocene" (presumed age of Aycross and/or Tepee Trail Formations and of some lower Wiggins stratal ages, respectively), to the Enos Creek-Owl Creek Debris-Avalanche (ECOCDA), even though that structure clearly involved principally the displacement of strata of the middle Eocene Tepee Trail and overlying Wiggins Formations. Given these stratigraphic constraints, any middle Eocene or later Eocene age for the ECOCDA would require the rapid, pre-debris-avalanche erosion of a minimum of 1,000 m of volcanic rock along the east front of the Absaroka Range—erosion necessary to form an escarpment for the debris-avalanche to have originated from, as well as to create the basin-ward void into which debris from the ECOCDA collapsed. There is absolutely no evidence for such a major erosional episode at a time when the Bighorn Basin adjacent on the east was filling with hundreds of meters of volcaniclastic sediment (*e.g.*, McKenna and Love, 1972). Moreover, there is also no evidence of any late Eocene (38.0-33.9 Ma) episode of the exceptionally violent volcanic activity required to trigger and sustain ECOCDA movement. Hiza (1999a) places the youngest Eocene Absaroka volcanic activity at ~43.95 Ma, well prior to the late Eocene (38.0-33.9 Ma).

The Enos Creek-Owl Creek Debris-Avalanche was most likely triggered by collapse of the Island Park Caldera at about 2.08 Ma (Christiansen, 2001; Swallow *et al.*, 2019), a series of exceptionally violent silicic "super eruptions" that resulted in deposition of the Huckleberry Ridge Tuff. The catastrophically displaced material was propelled rapidly eastward into the central Bighorn Basin atop a partially dissected, high-level, basinward-tilted erosion surface, a few scant remnants of which are preserved beneath allochthonous rocks at Adam Weiss Peak, Soapy Dale Peak, Leon Baird Peak, Noon Point, Squaw Buttes and, possibly, on several surfaces lacking allochthonous material such as Tatman Mountain and high points along the Meeteetse Divide.

Malone *et al.* (2014) considered the Tatman Formation of the central Bighorn Basin to be an equivalent of the basin margin Aycross Formation, and provided a date of 50.84 + 1.29/- 0.78 Ma for upper Tatman strata a few meters beneath the allochthonous rocks on West Squaw Butte (Table I). Tatman rocks in the central Bighorn Basin, including those at Squaw Buttes, elsewhere on the East Ridge Divide, and on Sheets Mountain and Tatman Mountain, contain little or no volcanic material, whereas the Aycross Formation is volcaniclastic. With the exception of some intertonguing rocks of Willwood lithology (variegated paleosols developed on floodplain mudstones) in some areas on Tatman Mountain and on the East Ridge Divide, Tatman rocks are of lacustrine origin. Carbonaceous shales and ostracodal paper shales are developed locally beneath the Aycross

Formation (Bown, 1982a), and tuffaceous lacustrine and lake margin strata interfinger with fluvial Aycross strata 50-70 meters above the base of the Aycross Formation in parts of the Enos Creek/Owl Creek area. Those lacustrine rocks are most likely equivalent to some part of the 250+ meters of volcanic detritus-bearing Eocene lacustrine strata in the Lysite Mountain area (atop the Owl Creek Mountains—Tourtelot, 1946; Wilson, 1964; Bay, 1969; Rohrer and Smith, 1969; Bown, 1982a), and not to the non-volcaniclastic Tatman rocks of the central Bighorn Basin.

Deposition of the preserved part of the Tatman Formation in the central Bighorn Basin clearly preceded Absaroka Range volcanism, whereas volcanic-rich Tatman-like rocks on Lysite Mountain resulted from a somewhat later stage of fluvio-lacustrine deposition (Rohrer and Smith, 1969). These stratigraphic relationships demonstrate that the allochthonous material capping Squaw Buttes did not slide out into a later early Eocene Tatman lake but, rather, it was propelled onto a much younger, early Pleistocene erosion surface developed on Tatman sediments.

Bown (1982a,b) concluded that only rocks of the Tepee Trail and Wiggins Formations were involved in the Enos Creek-Owl Creek debris-avalanche. The identification by Malone *et al.* (2014) of allochthonous Aycross-age rocks on Squaw Buttes demonstrates that Aycross rocks were indeed displaced during Enos Creek-Owl Creek debris-avalanching, and/or that significant volumes of Aycross rocks were *abducted/entrained* by the advancing mass (**Figure 54**), the effects of which would be the same. It is important to note, however, that Malone *et al.*'s (2014) date was obtained from a conglomerate *clast* and their date therefore merely shows that displacement of the rock took place at some indeterminate time after 50.53 + 1.59/- 0.78 Ma.

The Shoshone-Sunlight-Abiathar Detachment Fault (SSADF) displaced huge volumes of Paleozoic carbonate rocks as well as lesser amounts of volcanic and volcaniclastic rocks of the Cathedral Cliffs Formation. The Heart Mountain-McCullough Peaks Sturzstrom (HMMPS) displaced only Paleozoic carbonate rocks, none of which occur on Squaw Buttes. The otherwise farthest-flung volcanic remnants of the SSADF (*i.e.*, Cathedral Cliffs Formation or equivalents) are in the Shoshone River valley, at least 50 miles (>80 km) northwest of Squaw Buttes, and the displaced Paleozoic masses of the HMMPS closest to Squaw Buttes form the top of McCullough Peaks, some 45 miles (~74 km) to the north.

Between SSADF debris in the Shoshone River valley and Squaw Buttes lies the Meeteetse Rim at a maximum elevation of 6523 feet (~1989 m), 438 feet (~134 m) higher than the top of the erosion-truncated upper part of the Tatman Formation at West Squaw Butte. No displaced Paleozoic or volcanic material occurs anywhere on the Meeteetse Rim. Similarly, slightly east of the line joining McCullough Peaks and Squaw Buttes lie Sheets Mountain (elevation = 6295 feet = 1919 m) and Tatman Mountain (elevation 6261 feet = ~1909 m). No part of either rise is formed by allochthonous material of any kind. Elk Butte in the Oregon Basin rises to an elevation of 6210 feet (1893 m); the butte lies directly between the southeasternmost allochthons of the SSADF and Squaw Buttes yet exhibits no displaced Paleozoic carbonate or Eocene volcanic debris. These topographic relationships offer strong supporting, though not definitive, evidence that allochthonous rocks atop Squaw Buttes did not originate to the northwest.

The spatial/topographic relationship of the Squaw Buttes allochthons (red dot) and displaced rocks of the Enos Creek/Owl Creek Debris-avalanche is depicted in **Figure 55**. The Paleozoic carbonate rock masses on Heart Mountain and McCullough Peaks, believed by Malone *et al.* (2017) to be part of the same displacement as Squaw Buttes, lie well off the map, about 54 miles (~89 km) and 41 miles (~67 km) to the northwest and north-northwest, respectively.

Alleged Effects of the SSADF on Sedimentation in Lake Gosiute, Southwest Wyoming

Smith *et al.* (2014) proposed that during Farallon slab rollback from 50 to 47 Ma volcaniclastic sediments from magmatic centers were being deposited in Lake Gosiute in the Washakie Basin of southwestern Wyoming. The magmatic centers were coincident with uplifted regions of the overlying North American plate caused by dynamic thermal uplift from the influx of aesthenosphere above and adjacent to the descending slab. Rhodes *et al.* (2007) suggested that emplacement of the HMDF (= SSADF) upper plate in northwest Wyoming blocked a hypothetical river that flowed southward out of the Bighorn Basin and contributed volcaniclastic sediment to Lake Gosiute, an idea later endorsed by Malone and Craddock (2008). According to that interpretation, catastrophic damming of the hypothetical river by HMDF allochthons cut off its flow, causing the sudden desiccation of Lake Gosiute and resulting in the development of deep mudcracks on mudstones of the lower LaClede Bed of the Green River Formation. Atop the bed with mudcracks lies the "buff marker bed," a unit containing the first volcaniclastic sediments in the Washakie Basin and, by the Rhodes *et al.* (2007) scenario, volcaniclastics recording the first violent Absaroka Range volcanicity that immediately followed HMDF movement.

The Sand Butte Tuff in the upper part of the LaClede Bed was dated by O'Neill (1980) at 44.9 +/- 1.3 Ma to 45.2 +/- 1.7 Ma, giving the bed an age range of 46.9-43.5 Ma—much too young for it to be related in any way to volcanicity following emplacement of the HMDF (=SSADF) allochthons. Moreover, at the time of emplacement of SSADF allochthons (at approximately 49.5 Ma), the Tatman Lake occupied most of the central and southern Bighorn Basin and much of what is now the border area of the basin, and structural elevation of the anticlinal and the anticlinal and southwest- and south-thrust-faulted Washakie and Owl Creek ranges lay as effective barriers to any feeder drainage to Lake Gosiute that might have originated in the Bighorn Basin.

In their Figure 8, Malone *et al.* (2014) depict a paleodrainage running south, more-or-less down the structural axis of the Bighorn Basin (*i.e.*, through the Tatman Lake Basin), and abutting the "Washakie—Owl Creek Range." Similarly, earlier paleodrainage reconstructions offered by Surdam and Stanley (1980—used by Rhodes *et al.*, 2007) and by Craddock *et al.* (2015—essentially that of Malone *et al.*, 2014) inexplicably also cross the Tatman Lake Basin and/or (equally inexplicably) pass over several known elevated regions and structural highs.

Volcaniclastic sediment in Lake Gosiute was most likely derived from the 51-44 Ma Challis Volcanic Field (Moye *et al.*, 1988) *via* the "Idaho River," a drainage that entirely bypassed that part of the Bighorn Basin affected by the HMDF=SSADF (Chetel *et al.*, 2011). Therefore, there were no effects of the SSADF on the drying of Lake Gosiute because: 1) the SSADF and deposition of the LaClede Bed are separated by 2.6-5.9 Ma, and 2) there was no drainage southward, out of the Bighorn Basin that could have contributed to sedimentation in Lake Gosiute.

Finally, Lillegraven and Ostresh (1988:Figure 10) depict late Wasatchian—Bridgerian (= late early Eocene—early middle Eocene) Bighorn Basin streams draining into the Tatman Lake, not flowing southward toward the Green River Basin. Dunn (1979), cited in Surdam and Stanley (1980) and used by Rhodes *et al.* (2007) as evidence of stream confluence from the Bighorn to the Wind River Basin, deals with largely lacustrine middle-late Eocene rocks in the Lysite Mountain area that are considerably younger than the time of HMDF (SSADF) faulting. The most plausible reconstructions of Bighorn/Wind River drainage patterns near the time of HMDF movement are those of Hay (1956:Figure 10) and Lillegraven and Ostresh (1988:Figure 10). Paleocurrent studies of Seeland (1988) and Welch (2021) demonstrate that the only streams exiting the Bighorn Basin during (at least) the Willwood Formation part of the early Eocene did so to the north and northeast, and the controlling early Eocene stream in the Wind River Basin flowed to the east (Seeland, 1978).

Exotic Paleozoic Rocks Northeast of Carter Mountain

Exotic boulders and masses of jumbled, disaggregated Paleozoic carbonate debris not clearly associated with either SSADF faulting or the HMMPS deposit were first recognized in the Carter Mountain area by Dake (1918:49), who recorded:

> "... an isolated peak of Madison at the east end of Carter Mountain ..."
> and more Madison boulders several miles to the south, "... east and southeast of Carter Mountain."

Dake thought the exotics (*e.g.*, **Figure 56**) might betray a buried block of Paleozoic rocks; however, this seems unlikely as at least some of them lie clearly strewn across an eroded surface developed on the lower Eocene Willwood Formation, a setting similar to those at Heart Mountain and McCullough Peaks, some 21 miles (34.4 km) to the northeast and to the east-northeast, respectively.

Pierce (1968b: D237-D238) described Dake's "isolated peak of Madison" as:

> "... a jumbled mass of Paleozoic limestone and dolomite ¾ of a mile long and as much as 150 feet thick ... on landslide debris derived from the Willwood Formation of Eocene age ...".

Our studies show that dolomite boulders also occur farther to the west, where they are scattered across the divide west of Bridge Creek, suggesting that this allochthonous Paleozoic debris may once have extended over a somewhat larger area.

Pierce (1997) re-defined Dake's mass as a: "slumped block of limestone from upper plate of Heart Mtn. fault." Rocks directly underlying the mass itself are not exposed; however, an outcrop of the Willwood Formation only 300 m south of the Paleozoic debris (at 44° 21' 07.72" N, 109° 10' 43.26" W and environs) is well-exposed, flat-lying and clearly *in situ*, and is not backwards-rotated (as in a true slump), or deformed or reduced to rubble, as in a landslide deposit. Therefore, sitting as it does atop in place, undeformed Willwood rocks, the allochthonous Paleozoic carbonate debris at this site antedates and is unrelated to Pierce's (1968b) Carter Mountain landslides, which are demonstrably much younger.

In 1973, Pierce (his Fig. 3) placed the largest of the carbonate masses north of Carter Mountain on a map, considering it to have been (1973:465):

> "... *let down 2,100 feet* [640 m] *and 3 miles* [4.83 km] *horizontally on the Carter Mountain landslide in Quaternary time.*" (our brackets)

Our fieldwork indicates that the mass lies on undisturbed Willwood rocks and that movement probably originated to the north as part of the HMMPS or, less likely, as part of a separate, otherwise unknown large landslide deposit, the residue of which has been entirely eroded away. The architecture of the mass—mixed Paleozoic micritic and dolomitic blocks (up to 3 m diameter) lying atop comminuted carbonate debris—mirrors that of the Corbett masses on the west side of McCullough Peaks and the allochthonous Paleozoic material on the Peaks themselves. It is unlikely that the internal composition of this debris would survive intact after being "let down" the formidable vertical and horizontal distances suggested by Pierce, and there is no evidence of similar comminuted material outcropping *in situ* anywhere along the cliff faces of Carter Mountain.

Although debris of several relatively small, obviously much younger landslides and earthflows occurs in the area (*e.g.*, at 44° 31' 3" N, 109° 17' 18" W), a considerable volume of landslide material of an older, yet undetermined age (**Figure 33**) is preserved in the lower reaches of Rattlesnake Canyon—a narrow defile separating the allochthonous SSADF masses of Logan Mountain (on the west) from the Palisades area of Rattlesnake Mountain (on the east)—north of Buffalo Bill Reservoir. This landslide debris, not mapped as such by Nelson and Pierce (1968), but clearly visible in a large area surrounding and up-canyon from 44° 32' 6.69" N, 109° 17' 9.9" W, was derived from Paleozoic carbonates and it is interesting to speculate that it might have originated as a southern, outlying lobe of the HMMPS. The bearing of the Rattlesnake Canyon reach (approximately 140°) is such that the narrow confines of the steep canyon walls might have acted as a chute into which mobile HMMPS debris was funneled then propelled rapidly 26 miles (~43 km) down the canyon and across the ancient valley of the Shoshone River to the northeast edge of Carter Mountain. Such a mass would have dammed the Shoshone River at the Rattlesnake Mountain gap and to the south, re-directing its flow into and across the Oregon Basin (see below).

The source of this debris was probably at or near the junction of the northernmost part of Rattlesnake Mountain and the westernmost extremity of Pat O'Hara Mountain (**Figures 57 and 58**), an area exhibiting a large allochthon of Bighorn Dolomite and littered with lesser masses of unidentified Paleozoic carbonates lying atop erosionally truncated rocks of the Chugwater (Triassic) and Willwood (lower Eocene) Formations.

The top of the allochthonous carbonate mass northeast of Carter Mountain (at 44° 21' 19.48" N, 109° 10' 45.18" W) lies 446 meters above the bed of the adjacent South Fork of the Shoshone River (*e.g.*, at 44° 25' 55.22" N, 109° 15' 13.53" W). Using the formula determined by Palmquist (1978) to calculate the time of erosion from the top of the mass to the bed of the South Fork of the Shoshone River, the estimated age of emplacement of the allochthon is 2.350 Ma. The erosion rate formula determined by Love and Walsh (1981) gives an emplacement time of 2.194 Ma. These ages closely correspond to the 2.245 Ma and 2.091 Ma ages calculated for emplacement of the largest Corbett allochthon remnant, and lend credence to our suggestion that both allochthonous masses owe their origin to the Heart Mountain/McCullough Peaks Sturzstrom, as triggered by one of the exceedingly violent ~2.08 Ma Huckleberry Ridge Tuff eruptions.

Pierce and Andrews (1941) described carbonate boulders up to 2.5 feet (0.76 m) in diameter in their Cottonwood terrace southeast of Carter Mountain. The principal local drainages, Meeteetse Creek and Spring Creek and the controlling Greybull River, have no known outcrop sources for carbonate rocks. Pierce and Andrews also correlated the gravels of their "Rim" (= Meeteetse Rim) terrace with Mackin's (1937) Y-U Bench gravels, and their Sunshine terrace with Mackin's Emblem Bench. Both the "Rim" and Sunshine terrace gravels yield minor amounts of Paleozoic carbonate clasts, yet the

Greybull River has no Paleozoic outcrop source for that gravel. The highest (oldest) terrace gravel in the Bighorn Basin is the Fenton Pass Formation (Rohrer and Leopold, 1963), which caps truncated late early/early middle Eocene Tatman Formation sediments (Rohrer and Smith, 1969; Bown et al., 1994) atop Tatman Mountain, north of Squaw Buttes, in the west-central Bighorn Basin (see Table II). The Fenton Pass Formation is of early Pleistocene age, and its gravels include no Paleozoic carbonates.

Stevens (1938:1254) recorded the presence of limestone blocks up to 50 feet (15.2 m) in diameter entombed in "volcanic breccias" high above the base of the volcanic deposits on Carter Mountain. Stevens did not designate a specific locality for these breccias but his site almost certainly accords with the cliff face above Foster Reservoir (reservoir at, e.g., 44° 17' 03" N, 109° 11' 16" W). Stevens believed the exotic blocks were transported to their current positions by viscous Eocene breccia flows, similar to the situation he described west of Sheep Mountain, to the north, between the South and North Forks of the Shoshone River. Allochthonous carbonates in those flows might therefore be the source of some of the Paleozoic carbonate clasts in the Cottonwood, Meeteetse Rim, and Sunshine terraces. The breccias are not the source of the carbonate mass north of Carter Mountain originally described by Dake (1918) and discussed above, however, because that deposit lacks volcanic material and is composed of mostly comminuted carbonate debris that could not be transported intact.

∽ Roof Pendant Mass of Lower Paleozoic Rocks Capping Dollar Mountain

J.T. Rouse and E.H. Stevens (in Rouse, 1940:1424) published a 1,557-foot (475 m) measured stratigraphic section of exotic Paleozoic rocks lying astraddle Dollar Mountain in the high Absaroka Range above the head of the Wood River. Genetically related to the intrusion of a porphyritic quartz latite (Rouse, 1940), and/or a rhyolite stock (Wilson, 1964), these exotic rocks comprise a massive roof pendant (Antweiler et al., 1985) made up of Cambrian through Pennsylvanian carbonates, shales, and sandstones nearly one mile^2 (1.61 km^2) in area. According to Rouse (1940), rocks exposed include: the Cambrian Flathead, Gros Ventre, and Gallatin formations; the Ordovician Bighorn Formation; the Devonian Darby Formation; the Mississippian Madison Formation; and part of the Pennsylvanian Amsden Formation. The top of the 1.23 km^3 pendant mass dips 25°-45° to the north-northwest and lies at least 6,546 feet (1,995 m) above the top of stratigraphically equivalent, in situ Paleozoics encountered in a borehole on the Wood River 5.5 miles (8.68 km) east-southeast of Dollar Mountain (Wilson, 1982), and 3,485 feet (1,063 m) above the top of the nearest exposed in situ Paleozoic rocks, which outcrop seven miles (~11.3 km) south-southeast of Snow Lake, some 10 miles (16.1 km) southwest of Dollar Mountain (Love and Christiansen, 1985).

Though clearly associated with violent volcanic activity of some duration, the intrusion that elevated the roof pendant mass of Paleozoic strata forms part of the mineralized body of the Kirwin intrusive complex (e.g., Hewett, 1912; Wilson, 1964, 1971, 1975b; Hausel, 1982). Those intrusions invaded rocks of the middle and upper Eocene Wiggins Formation and therefore clearly post-date late early Eocene Shoshone/Sunlight/Abiathar (SSADF) detachment faulting. The Paleozoic rocks atop Dollar Mountain do, however, constitute a likely source for some of the Paleozoic carbonate clasts occurring in the Cottonwood, Meeteetse Rim, and Sunshine terraces discussed above, because those terraces lie below the confluence of the Wood River with the Greybull River.

ACKNOWLEDGMENTS

The authors' interest in the Heart Mountain Detachment Fault problem began in 1967 when Al and Tom speculated that Corbett Mass #2 was of landslide origin. That prospect was examined further by Bown and Love (1987). Al Warner began a photographic survey of the HMDF in 2015, and Al and Tom rekindled our interest in all aspects of the HMDF problem and began a joint program of field research in 2017, adding Mark Mathison to our team in 2024.

Access to important outcrops has been curtailed considerably in the last few decades by the closure of many important roads to public access, including several that appear to provide the only admission to outliers of Bureau of Land Management or U.S. Forest Service lands. We are most indebted to Google Earth for satellite imagery access to all parts of the study area, and Google Earth is additionally acknowledged in each of the pertinent figure captions.

We are supremely grateful to Winston Churchill (Powell, WY), Larry Luckenbill (Cooke City, MT), and Brad Sorenson (Wyoming Game and Fish Commission, Cody, WY) for valuable assistance in accessing field areas, and to Terri Briggs (Big Moose Lodge, Cooke City, MT) and Matt Dzialak and Jenine Phelps (DDX Ranch, Cody, WY) for consultation regarding access to important geologic areas and for their provident hospitality. We enjoyed a highly informative reconnaissance of the north side of Heart Mountain thanks to Heart Mountain Ranch Manager Brian Peters, and we are grateful to Jenette of the Moon Crest Ranch for permission to explore Rattlesnake Canyon.

We thank Archway editors Aimee Reff and Bob DeGroff for their assistance with all aspects of publication, and are grateful for technical comments on the manuscript by geologists Dr. Will Clyde (University of New Hampshire), Dr. Gary Johnson (Dartmouth University), and Dr. Erik Kvale (Shell, WY). Additionally, we thank Ben Rodwell (University of Texas at Austin) and Drs. Amy Chew (Brown University), Jeff Eaton (Weber State University, retired), J. David Love (U.S. Geological Survey—deceased), Ingrid Lundeen (Hunter College, NY), William G. Pierce (U.S. Geological Survey—deceased), and Carl F. Vondra (Iowa State University, emeritus) for geological discussion. M. Craig Campbell (Big Horn, WY—deceased) and Andrew McKenna (Boulder, CO) were T. Bown's able field assistants in 1977 and 1979, respectively. J. Colter Johnson (Fort Collins, CO) was our driver, field assistant, and security against bears in 2021. We are grateful to Dr. Dan Hummer (Southern Illinois University at Carbondale, IL) who analyzed and interpreted samples of Paleozoic carbonate breccias. Kimberly Nichols, Dr. Sayat Tamirbekov, Rachel Winter, and Jason Kapernekas (Colorado State University) and Kat Andersen (Fort Collins, CO) assisted considerably with editing and preparation of the illustrations.

This study is dedicated to Dr. William G. Pierce (1904-1994; **Figure 59**) and Dr. J. David Love (1913-2002; **Figure 60**), two titans of Absaroka Range geology, in acknowledgment of their friendship and generous consultation with T. Bown over many years, and their numerous outstanding contributions to our understanding of Wyoming geology, and to Dr. Carl F. Vondra (**Figure 61**), world-class field geologist, our mentor, and Director of Iowa State University's Carl F. Vondra Geology Field Station from 1965 to 2003. All three authors are proud alumni of Iowa State University and of several seasons each at the Geology Field Station.

All photographic images in this report were taken by A. J. Warner, unless otherwise indicated. Figure 1 was modified with permission from Figure 1 of Lundeen and Kirk (2023).

REFERENCES CITED

Abele, G. 1994. Large rockslides: their causes and movement on internal sliding planes. Mountain Research and Development, 14:315-320.

Abrahams, A.D., Parsons, A.J., Cooke, R.U., and Reeves, R.W. 1984. Stone movement on hillslopes in the Mojave Desert, California: A 16-year record. Earth Surface Processes and Landforms, 9:365-370.

Aharonov, E., and Anders, M.H. 2006. Hot water: A solution to the Heart Mountain detachment problem? Geology, 34:165-168.

Ambrose, S.H. 1998. Late Pleistocene human population bottlenecks, volcanic winter, and differentiation of modern humans. Journal of Human Evolution, 34:623-651.

Anders, M.H., Fouke, B.W., Zerkle, A.L., Tavarnelli, E., Alvarez, W., and Harlow, G.E. 2010. The role of calcining and basal fluidization in the long runout of carbonate slides: An example from the Heart Mountain slide block, Wyoming and Montana, U.S.A. Journal of Geology, 118:577-599.

Antweiler, J.C., Rankin, D.W., Fisher, F.S., Long, C.L., Love, J.D., Bieniewski, C.L., and Smith, R.C. 1985. Mineral resource potential of the northern part of the Washakie Wilderness and nearby roadless areas, Park County, Wyoming. U.S. Geological Survey Miscellaneous Field Studies Map MF-1597-A and pamphlet, pp. 1-24.

Applegarth, M.T. 2004. Assessing the influence of mountain slope morphology on pediment form, south-central Arizona. Physical Geography, 25:226-236.

Barrientos, S.E., and Ward, S.N. 1990. The 1960 Chile earthquake: inversion for slip distribution from surface deformation. International Geophysical Journal, 103:589-598.

Bay, K.W. 1969. Stratigraphy of Eocene sedimentary rocks in the Lysite Mountain area, Hot Springs, Fremont, and Washakie counties, Wyoming. Laramie, University of Wyoming unpublished PhD Thesis:181 pp.

Beutner, E.C. 2002. Accreted grains as the "smoking gun" of the Heart Mountain Detachment Fault, NW Wyoming. Geological Society of America Abstracts with Programs, Paper 165-1.

Beutner, E.C., and Craven, A.E. 1996. Volcanic fluidization and the Heart Mountain detachment, Wyoming. Geology, 24:595-598.

Beutner, E.C., and DiBenedetto, S.P. 2003. The Blacktail thrust-fold, Crandall Conglomerate, and Heart Mountain Detachment Fault, northwestern Wyoming. Rocky Mountain Geology, 38:237-245.

Beutner, E.C., and Gerbi, G.P. 2005. Catastrophic emplacement of the Heart Mountain block slide, Wyoming and Montana, USA. Geological Society of America Bulletin, 117:724-735.

Beutner, E.C., and Hauge, T.A. 2009. Heart Mountain and South Fork fault systems: Architecture and evolution of the collapse of an Eocene volcanic system, northwest Wyoming. Rocky Mountain Geology, 44:147-164.

Bilham, R., and England, P. 2001. Plateau "pop-up" in the great 1897 Assam earthquake. Nature, 410:806-809.

Bletery, Q., Thomas, A., Rempel, A.W., Karlstrom, L., Sladen, A., and DeBarros, L. 2016a. Fault curvature may control where big quakes occur. Eurekalert, 24 November, 2016.

Bletery, Q., Thomas, A., Rempel, A.W., Karlstrom, L., Sladen, A., and DeBarros, L. 2016b. Mega-earthquakes rupture flat megathrusts. Science, 354:1027-1031.

Bosai, K. 2011. About strong ground motion caused by the 2011 earthquake off the coast of Tohoku. Natur. 410.806B.

Bourne, J.A., and Twidale, C.R. 1998. Pediments and alluvial fans: Genesis and relationships in the western piedmont of the Flinders Ranges, South Australia. Australian Journal of Earth Sciences, 45:123-135.

Bown, T.M. 1982a. Geology, paleontology, and correlation of Eocene volcaniclastic rocks, southeast Absaroka Range, Hot Springs County, Wyoming. U.S. Geological Survey Professional Paper 1201-A: 75 pp.

Bown, T.M. 1982b. Catastrophic large-scale Late Cenozoic detachment faulting of Eocene volcanic rocks, southeast Absaroka Range, northwest Wyoming. Wyoming Geological Association 33rd Annual Field Conference Guidebook:185-201.

Bown, T.M., Chew, A.E., Nichols, K.A., Rose, K.D., Rodwell, B.W., and Weaver, L.N. 2016. Paleogene sedimentary tectonics of the Greybull/Basin area, northwest Wyoming: Little Dry Creek parafold and Gould Butte/Three Sisters block. Geological Society of America, Abstracts with Programs, Paper 347-24.

Bown, T.M., and Larriestra, C.N. 1990. Sedimentary paleoenvironments of fossil platyrrhine localities, Miocene Pinturas Formation, Santa Cruz Province, Argentina. Journal of Human Evolution, 19:87-119.

Bown, T.M., and Love, J.D. 1985. Comment and Reply on "Catastrophic debris avalanche from ancestral Mount Shasta volcano, California". Geology, 13:79-80.

Bown, T.M., and Love, J.D. 1987. The Rhodes allochthon of the Enos Creek-Owl Creek debris-avalanche, northwest Wyoming. Geological Society of America Centennial Field Guide—Rocky Mountain Section:179-182.

Bown, T.M., and Love, J.D. 1989. Possible Late Cenozoic emplacement of Paleozoic allochthons on Heart Mountain and McCulloch Peaks, Park County, Wyoming. Geological Society of America Abstracts with Programs, 21:A61.

Bown, T.M., Rose, K.D., Simons, E.L., and Wing, S.L. 1994. Distribution and stratigraphic correlation of upper Paleocene and lower Eocene fossil mammal and plant localities of the Fort Union, Willwood, and Tatman Formations, southern Bighorn Basin, Wyoming. U.S. Geological Survey Professional Paper 1540:1-103.

Bozorgnia, Y., and Campbell, K. 2016. Ground motion model for the vertical-to-horizontal (V/H) ratios of PGA, PGV, and response spectra. Earthquake Spectra, 32:951-978.

Branney, M., and Acocella, V. 2015. Calderas. The Encyclopedia of Volcanoes (Second Edition):299-315.

Bucher, W.H. 1933. Volcanic explosions and overthrusts. Transactions of the American Geophysical Union, Reports and Papers, Volcanology, p. 238-242.

Bucher, W.H. 1947. Heart Mountain Problem. University of Wyoming, Wyoming Geological Association, and Yellowstone-Bighorn Research Association 1947 Field Conference Guidebook, pp.189-197.

Camp, V.E., and Wells, R.E. 2021. The case for a long-lived and robust Yellowstone Hotspot. GSA Today, 31:1-4.

Chadwick, R.A. 1970. Belts of eruptive centers in the Absaroka—Gallatin Volcanic Province, Wyoming and Montana. Geological Society of America Bulletin, 81:267-274.

Chesner, C.A., and Rose, W.I. 1991. Stratigraphy of the Tuba Tuffs and the evolution of the Tuba Caldera complex, Sumatra, Indonesia. Bulletin of Volcanology, 54:343-356.

Chetel, L.M., Janecke, S.U., Carroll, A.R., Beard, B.J., Johnson, C.M., and Singer, B.S. 2011. Paleogeographic reconstruction of the Eocene Idaho River, North American Cordillera. Geological Society of America Bulletin, 123:71-88.

Chew, A.E. 2005. Biostratigraphy, paleoecology, and synchronized evolution in the early Eocene mammalian fauna of the central Bighorn Basin, Wyoming. PhD Dissertation, Baltimore, The Johns Hopkins University, 660 pp.

Christiansen, R.L. 2001. The Quaternary and Pliocene Yellowstone Plateau volcanic field of Wyoming, Idaho, and Montana. U.S. Geological Survey Professional Paper, 729-G:1-145.

Christiansen, R.L., and Blank, H.R., Jr. 1972. Volcanic stratigraphy of the Quaternary rhyolite plateau in Yellowstone National Park; pp. B1-B18 *in* Geology of Yellowstone National Park, U.S. Geological Survey Professional Paper 729-B.

Christiansen, R.L., Foulger, G.R., and Evans, J.R. 2002. Upper-mantle origin of the Yellowstone hot spot. Geological Society of America Bulletin, 114:1245-1256.

Clyde, W.C. 2001. Mammalian biostratigraphy of the McCullough Peaks area in the northern Bighorn Basin; pp. 109-126 *in* P.D. Gingerich (ed.), Paleocene-Eocene Stratigraphy and Biotic Change in the Bighorn and Clark's Fork Basins, Wyoming. University of Michigan Papers on Paleontology, no. 33:109-126.

Colgan, J.P. 1998. Heart Mountain faulting and the emplacement of intrusive rocks at Painter Gulch, near White Mountain, northwest WY. Proceedings of the Keck Research Symposium on Geology, 11: 299-302.

Craddock, J.P., Geary, J., and Malone, D.H. 2012. Vertical injectites of detachment carbonate ultracataclasite at White Mountain, Heart Mountain detachment, Wyoming. Geology, 40:463-466.

Craddock, J.P., Malone, D.H., Magloughlin, J., Cook, A.L., Riesner, M.E., and Doyle, J.R. 2009. Dynamics of the emplacement of the Heart Mountain allochthon at White Mountain: Constraints from calcite twinning strains, anisotrophy of magnetic susceptibility, and thermodynamic calculations. Geological Society of America Bulletin, 121:919-938.

Craddock, J.P., Malone, D.H., Porter, R., MacNamee, A., Mathisen, M., Kravitz, K., and Leonard, A. 2015. Structure, timing, and kinematics of the early Eocene South Fork Slide, northwest Wyoming, USA. Journal of Geology, 123:311-335.

Craddock, J.P., Nielson, K.J., and Malone, D.H. 2000. Calcite twinning strain constraints on the emplacement rate and kinematic pattern of the upper plate of the Heart Mountain Detachment. Journal of Structural Geology, 22:983-991.

Crandell, D.R., Miller, C.D., Glicken, H.X., Christiansen, R.L., and Newhall, C.G. 1984. Catastrophic debris avalanche from ancestral Mount Shasta volcano, California. Geology, 12:143-146.

Crozier, M.J. 1992. Determination of paleoseismicity from landslides; pp. 1173-1180 *in* Bell, D.H. (ed.),

Landslides (Glissements de terrain), vol 2. Rotterdam, A.A. Balkema.

Cruden, D.M., and Hungr, O. 1986. The debris of the Frank Slide and theories of rockslide—avalanche mobility. Canadian Journal of Earth Sciences, 23:425-432.

Currie, C.A., and Copeland, P. 2022. Numerical models of Farallon Plate subduction: creating and removing a flat slab. Geosphere, https://doi.org/10. 1130/GES02393. 1.

Dade, W.B., and Huppert, H.E. 1998. Long-runout rockfalls. Geology, 26:803-806.

Dahren, B., Troll, V.R., Andersson, U.B., Chadwick, J.P., Gardner, M.F., Jaxybulatov, K., and Koulakov, I. 2012. Magma plumbing beneath Anak Krakatau volcano, Indonesia: Evidence for multiple magma storage regions. Contributions to Mineralogy and Petrology, 163:631-651.

Dake, C.L. 1918. The Hart Mountain overthrust and associated structures in Park County, Wyoming. Journal of Geology, 26:45-55.

Davis, G.A. 1965. Role of fluid pressure in mechanics of overthrust faulting: Discussion. Geological Society of America Bulletin, 76:463-468.

Davis, W.M. 1930. Rock floors in arid and in humid climates. Journal of Geology, 38:1-27.

Davies, T.R., McSaveney, M.J., and Hodgson, K.A. 1999. A fragmentation-spreading model for long-runout rock avalanches. Canadian Journal of Geotechnology, 36:1096-1110.

Decker, P.L. 1990. Style and mechanics of liquefaction-related deformation, lower Absaroka Volcanic Supergroup (Eocene) Wyoming. Geological Society of America Special Paper 240:71 pp.

Degenkolb, H.J. 1971. Preliminary structural lessons from the earthquake; pp. 133-134 in The San Fernando, California, earthquake of February 9, 1971, U.S. Geological Survey Professional Paper 733.

Douglas, T.A., Chamberlain, C.P., Poage, M.A., Abruzzere, M., Schultz, S., Henneberry, J., and Layer, P. 2003. Fluid flow and the Heart Mountain fault: A stable isotopic, fluid inclusion, and geochronologic study. Geofluids, 3:13-32.

Dunn, T.L. 1979. Mineral reactions in sandstones of the Lysite Mountain area, central Wyoming. Laramie, M.Sci. Thesis, The University of Wyoming:1-44.

Eldridge, G.H. 1894. A geological reconnaissance in northwest Wyoming. U.S. Geological Survey Bulletin, 119:30-31.

Encyclopedia Britannica. 2023. Mount Tambora Volcano, Indonesia.

Fauque, L., and Strecker, M.R. 1988. Large rock avalanches (Sturzstrüme, sturzstroms) at Sierra Aconquija, northern Sierras Pampeanas, Argentina. Eclogae geologicae Helvetiae, 81:579-592.

Feeley, T.C., and Cosca, M.A. 2003. Time vs. composition trends of magmatism at Sunlight volcano, Absaroka volcanic province, Wyoming. Geological Society of America Bulletin, 115:714-728.

Fisher, C.A. 1906. Geology and water resources of the Bighorn Basin, Wyoming. U.S. Geological Survey Professional Paper, 53:1-72.

Foden, J., and Varne, R. 1980. The petrology and tectonic setting of Quaternary-Recent volcanic centres of Lombok and Sumbawa, Sunda Arc. Chemical Geology, 30:201-206.

Francis, P.W., and Self, S. 1987. Collapsing volcanoes. Scientific American, 278:90-99.

Fritz, W.J., and Thomas, R.C. 2011. Roadside Geology of Yellowstone Country. Missoula, MT; Mountain Press: 309 pp.

Gerbi, G. 1997. The breccia layer of the Heart Mountain detachment and related clastic dikes. Proceedings of the Keck Research Symposium on Geology, 10:146-149.

Gibbard, P., and Head, M.J. 2009. IUGS ratification of the Quaternary System/Period and the Pleistocene Series/Epoch with a base at 2.58 Ma. Quaternaire, 20:411-412.

Giegengack, R., Omar, G.I., and Johnson, K.R. 1986. A reconnaissance fission track uplift chronology for the northwest margin of the Bighorn Basin; pp. 179-184 in Geology of the Beartooth Uplift and Adjacent Basins, Montana. Montana Geological Society and Yellowstone-Bighorn Research Association Joint Field Conference Guidebook.

Giegengack, R., Omar, G.I., Lutz, T.M., and Johnson, K.R. 1988. Tectono-thermal history of the Bighorn Basin and adjacent highlands, Montana-Wyoming. Geological Society of America Abstracts with Programs, 20 (6):416.

Gingerich, P.D. 1980. Evolutionary patterns in Early Cenozoic mammals. Annual Reviews of Earth and Planetary Sciences, 8:407-424.

Gingerich, P.D. 1991. Systematics and evolution of early Eocene Perissodactyla (Mammalia) in the Clark's Fork Basin, Wyoming. Contributions from the Museum of Paleontology of the University of Michigan, 28:181-213.

Glen, J.M.G., and Ponce, D.A. 2002. Large-scale fractures related to inception of the Yellowstone hot spot. Geology, 30:647-650.

Goguel, J. 1969. Le rôle de l'eau et de la chaleur dans les phénomènes tectoniques. Revue Géographie Physiographie et Géologie Dynamique, 11:153-164.

Goguel, J. 1978. Scale-dependent rockslide mechanisms, with emphasis on the role of pore fluid vaporization; pp. 693-705 in B. Voight (ed.), Rockslides and Avalanches. Amsterdam, Elsevier.

Goren, L., Aharonov, E., and Anders, M.H. 2010. The long runout of the Heart Mountain landslide: Heating, pressurization, and carbonate decomposition. Journal of Geophysical Research, 115:B10210.

Götze, J. 2009. Chemistry, textures, and physical properties of quartz—geological interpretation and technical application. Mineralogical Magazine, 73:645-671.

Greshko, M. 2016. 201 years ago this volcano caused a climate catastrophe. National Geographic Newsletter: April 8, 2016.

Hadley, R.F. 1967. Pediments and pediment-forming processes. Journal of Geological Education, 15:83-89.

Haeussler, P.J., Bradley, D.C., Wells, R.E., and Miller, M.L. 2003. Life and death of the Resurrection Plate: evidence for its existence and subduction in the northeastern Pacific in Paleocene-Eocene time. Geological Society of America Bulletin, 115:867-880.

Hague, A. 1899. Description of the Absaroka Quadrangle (Crandall and Ishawooa Quadrangles). U.S. Geological Survey Geological Atlas, Absaroka Folio, no. 52:1-6.

Hares, C.J. 1934. Relative age of the Heart Mountain overthrust and the Yellowstone Park volcanic series. Geological Society of America Proceedings for 1933, pp. 84-85.

Harrison, J.V., and Falcon, N.L. 1938. An ancient landslip at Saidmarreh in southwestern Iran. Journal of Geology, 46:296-309.

Hauge, T.A. 1982. The Heart Mountain Detachment Fault, northwest Wyoming: involvement of Absaroka volcanic rock. Wyoming Geological Association 33rd Annual Field Conference Guidebook:175-180.

Hauge, T.A. 1983. Geometry and kinematics of the Heart Mountain detachment fault, northwestern Wyoming and Montana. University of Southern California PhD Thesis, 267 pp.

Hauge, T.A. 1985. Gravity-spreading origin of the Heart Mountain allochthon, northwestern Wyoming. Geological Society of America Bulletin, 96:1440-1456.

Hauge, T.A. 1990a. The case for a continuous Heart Mountain allochthon. Wyoming Geological Association 41st Annual Field Conference Guidebook:183-185.

Hauge, T.A., 1990b. Kinematic model of a continuous Heart Mountain allochthon. Geological Society of America Bulletin, 102:1174-1188.

Hauge, T.A. 1993. The Heart Mountain detachment, northwestern Wyoming: 100 years of controversy; pp. 530-571 in Snoke, A.W., Steidtmann, J.R., and Roberts, S.M. (eds.), *Geology of Wyoming* (Blackstone-Love Volume). Wyoming Geological Survey Memoir #5.

Hausel, W.D. 1982.General geologic setting and mineralization of the porphyry copper deposits, Absaroka Volcanic Plateau, Wyoming. Wyoming Geological Association 33rd Annual Field Conference Guidebook:297-313.

Hay, R.L. 1954. Structural relationships of tuff-breccia in Absaroka Range, Wyoming. Geological Society of America Bulletin, 65:605-620.

Hay, R.L., 1956. Pitchfork Formation, detrital facies of Early Basic Breccia, Absaroka Range, Wyoming. American Association of Petroleum Geologists Bulletin, 40:1863-1898.

Hayes, G.P., Myers, E.K., Dewey, J.W., Briggs, R.W., Earle, P.S., Benz, H.M., Smoczyk, G.M., Flamme, H.E., Barnhart, W.D., Gold, R.D., and Furlong, K.P. 2017. Tectonic summaries of magnitude 7 and greater earthquakes from 2000 to 2015. U.S. Geological Survey Open-file Report 2016-1192:1-148.

Heasler, H.P., Jaworowski, C. Jones, R.W., De Bruin, R.H., and Ver Ploeg A.J. 1996. A self-guided geologic tour of the Chief Joseph Scenic Highway and surrounding area, northwestern Wyoming. Wyoming State Geological Survey, Public Information Circular No. 35.

Heim, A. 1882. Der Bergstürz von Elm. Deutsch. Geologischichte Gesellschaft Zeitschrift, 34:74-115.

Heim, A. 1883. Der alte Bergstürz von Flims. Alpenclubs, 18:295-309.

Heim, A. 1932. *Bergstürz und Menschenleben*. Zurich, Fretz und Wasmuth, 218 pp.

Hendrix, M.F. 2011. Geology Underfoot in Yellowstone Country. Missoula, MT; Mountain Press: 301 pp.

Hewett, D.F. 1912. The ore deposits of Kirwin, Wyoming. U.S. Geological Survey Bulletin, 540:121-132.

Hewett, D.F. 1920. The Heart Mountain overthrust, Wyoming. Journal of Geology, 28:536-557.

Hewett, D.F. 1926. Geology and oil and coal resources of the Oregon Basin, Meeteetse, and Grass Creek Basin quadrangles, Wyoming. U.S. Geological Survey Professional Paper, 145:1-111.

Hewitt, K. 1988. Catastrophic landslide deposits in the Karakoram Himalaya. Science, 242:64-67.

Hiza, M.M. 1999a. The geochemistry and geochronology of the Eocene Absaroka Volcanic Province, northern Wyoming and southwest Montana, U.S.A. Oregon State University PhD Thesis.

Hiza, M.M. 1999b. Protracted deformation (>2 Ma) of the Heart Mountain detachment, Absaroka Volcanic Province, Wyoming. Geological Society of America Abstracts with Programs, 31:428.

Hooper, P., Camp, V.E., Reidel, S.P., and Ross, M.E. 2007. The origin of the Columbia River flood basalt province: Plume versus nonplume models; pp. 635-668 in Foulger, G.R. and Jurdy, D.M. (eds.), Plates, Plumes, and Planetary Processes. Geological Society of America Special Paper 430.

Hoppin, R.A. 1974. Lineaments: their role in tectonics of central Rocky Mountains. American Association of Petroleum Geologists Bulletin, 58:2260-2273.

Hoppin, R.A., and Jennings, T.V. 1971. Cenozoic tectonic elements, Bighorn Mountain region, Wyoming-Montana; pp. 39-47 in Wyoming Geological Association 23rd Annual Field Conference Guidebook.

Hsü, K.J. 1969. Role of cohesive strength in the mechanics of overthrust faulting and of landsliding. Geological Society of America Bulletin, 80:927-952.

Hsü, K.J. 1975. Catastrophic debris streams (Sturzstroms) generated by rockfalls. Geological Society of America Bulletin, 86:129-140.

Hsü, K.J. 1978. Albert Heim: Observations on landslides and relevance to modern interpretations; pp. 70-93 in Voight, B. (ed.), Rockslides and Avalanches, 1. Natural Phenomena. New York, Elsevier.

Hubbert, M.K., and Rubey, W.W. 1959. Role of fluid pressure in mechanics of overthrust faulting. I. Mechanics of fluid-filled porous solids and its applications to overthrust faulting. Geological Society of America Bulletin, 70:115-166.

Hughes, C.J. 1970. The Heart Mountain detachment fault—A volcanic phenomenon? The Journal of Geology, 78:107-116.

Izett, G.A., and Wilcox, R.E. 1982. Map showing localities and inferred distributions of the Huckleberry Ridge, Mesa Falls, and Lava Creek ash beds (Pearlette Family ash beds) of Pliocene and Pleistocene age in the western United States and southern Canada. U.S. Geological Survey Miscellaneous Investigations Map, I-1325, scale 1:400,000.

Jibson, R.W. 1996. Use of landslide for paleoseismic analysis. Engineering Geology, 43:291-323.

Johnson, D. 1885. Planes of lateral corrasion. Science, 73:174-177.

Johnson, D. 1932. Rock fans of arid regions. American Journal of Science, 23:389-416.

Keefer, D.K. 1984. Landslides caused by earthquakes. Geological Society of America Bulletin, 95:406-421.

Kehle, R.O. 1970. Analysis of gravity sliding and orogenic translation. Geological Society of America Bulletin, 81:1641-1664.

Kent, P.E. 1965. The transport mechanism in catastrophic rock falls. The Journal of Geology, 74:79-83.

King, E.M., Malone, D.H., and DeFrates, J. 2009. Oxygen isotope evidence of heated pore fluid interaction with mafic dikes at Cathedral Cliffs, Wyoming. Mountain Geologist, 46:1-18.

King, L.C. 1949. The pediment landform: Some current problems. Geological Magazine, 86:245-250.

Korzec, A. 2016. The effect of the vertical acceleration on stability assessment of seismically loaded earth dams. Archives of Hydro-engineering and Environmental Mechanics, 63:101-120.

Kraus, M.J. 1992. Alluvial response to differential subsidence: sedimentological analysis aided by remote sensing, Willwood Formation (Eocene) Bighorn Basin, Wyoming, USA. Sedimentology, 39:455-470.

Lane, C.S., Chorn, B.T., and Johnson, T.C. 2013. Ash from the Toba supereruption in Lake Malawi shows no volcanic winter in East Africa at 75 Ka. Proceedings of the National Academy of Sciences, 110:8025-8029.

Lay, T., Kanamori, H., Ammon, C., Nettles, M., Ward, S.N., Aster, R., Beck, S.L., Bilek, S., Brudzinski, R., DeShon, H.R., Ekstrom, G., Sataki, K., and Sipkin, S. 2005. The great Sumatran-Andaman earthquake of 26 December 2014. Science, 308:1127-1133.

Leopold, L.B., Emmett, W.W., and Myrick, R.M. 1966. Channel and hillslope processes in a semiarid area, New Mexico. U.S. Geological Survey Professional Paper, 352-G:193-252.

Le Page, Y., and Donnay, G. 1976. Refinement of the crystal structure of low-quartz. Acta Crystallographica, B32:2456.

Lillegraven, J.A. 2009. Where was the western margin of northwestern Wyoming's Bighorn Basin late in the early Eocene? Pp. 37-81 *in* L.E. Albright III (ed.), Papers on Geology, Vertebrate Paleontology, and Biostratigraphy in Honor of Michael O. Woodburne. Bulletin of the Museum of Northern Arizona, 65.

Lillegraven, J.A., and Ostresh, L.M., Jr. 1988. Evolution of Wyoming's Early Cenozoic topography and drainage patterns. National Geographic Research, 4:303-327.

Liu, L., and Stegman, D.R. 2012. Origin of Columbia River flood basalt controlled by propagating rupture of the Farallon slab. Nature, 482:386-389.

Losh, S. 2018. Block sliding, Heart Mountain detachment, Wyoming. Geofluids, 3: 13-32.

Love, J.D. 1939. Geology along the southern margin of the Absaroka Range, Wyoming. Geological Society of America Special Paper, 20:1-133.

Love, J.D. 1960. Cenozoic sedimentation and crustal movement in Wyoming. American Journal of Science, 258-A:204-214.

Love, J.D., and Christiansen, A.C. 1985. Geologic Map of Wyoming. U.S. Geological Survey Map; 1:500,000 (3 sheets).

Love, J.D., McKenna, M.C., and Dawson, M.R. 1976. Eocene, Oligocene, and Miocene rocks and vertebrate fossils at the Emerald Lake locality, 3 miles south of Yellowstone National Park, Wyoming. U.S. Geological Survey Professional Paper, 932-A:1-28.

Love, J.D., and Walsh, H. 1981. Personal written communication by J.D. Love to T. Bown.

Lundeen, I.K., and Kirk, E.C. 2023. Euarchontans from Fantasia, an upland middle Eocene locality at the western margin of the Bighorn Basin. Journal of Human Evolution, 176 (2023) 103310.

Mackin, J.H. 1936. The capture of the Greybull River. American Journal of Science, 31:373-385.

Mackin, J.H. 1937. Erosional history of the Big Horn Basin, Wyoming. Geological Society of America Bulletin, 48:813-894.

Mackin, J.H. 1947. Altitude and local relief of the Bighorn area during the Cenozoic. University of Wyoming, Wyoming Geological Association, and Yellowstone-Bighorn Research Association Field Conference in the Bighorn Basin Guidebook:103-120.

Madsen, J.K., Thorkelson, D.J., Friedman, R.M., and Marshall, D.D. 2006. Cenozoic to Recent plate configurations in the Pacific Basin: Ridge subduction and slab window magmatism in western North America. Geosphere, 2:11-34.

Maley, R.P., and Cloud, W.K. 1971. Preliminary strong-motion results from the San Fernando earthquake of February 9, 1971; pp. 163-176 in The San Fernando, California, earthquake of February 9, 1971, U.S. Geological Survey Professional Paper 733.

Malone, D.H. 1995. Very large debris-avalanche deposit within the Eocene volcanic succession of the northeastern Absaroka Range, Wyoming. Geology, 23:661-664.

Malone, D.H. 2000. Gigantic Eocene landslide/debris-avalanche deposits of Wyoming's eastern Absaroka Range. Geological Society of America Abstracts with Programs, 32:A79.

Malone, D.H., Breeden, J.R., Craddock, J.P., Anders, M.H., and Macnamee, A. 2015. Age and provenance of the Eocene Crandall Conglomerate: Implications for the emplacement of the Heart Mountain slide. Mountain Geologist, October 2015:241-269.

Malone, D.H., and Craddock, J.P. 2008. Recent contributions to the understanding of the Heart Mountain Detachment, Wyoming. Northwest Geology, 37:21-40.

Malone, D.H., Craddock, J.P., Anders, M.H., and Wulff, A. 2014. Constraints on the emplacement age of the Heart Mountain slide, northwestern Wyoming. The Journal of Geology, 122:671-686.

Malone, D.H., Craddock, J.P., Garber, K.L., and Trela, J. 2017. Detrital zircon geochronology of the Aycross Formation (Eocene) near Togwotee Pass, western Wind River Basin, Wyoming. The Mountain Geologist, 54:69-85.

Malone, D.H., Craddock, J.P., and Mathesin, M.G. 2014. Origin of allochthonous volcanic rocks at Squaw Peaks, Wyoming: A distal remnant of the Heart Mountain slide? The Mountain Geologist, 51:329-344.

Malone, D.H., Craddock, J.P., Schmitz, M.D., Kenderes, M., Kraushaar, B., Murphey, C.J., Nielson, S., and Mitchell, T.M. 2017. Volcanic initiation of the Eocene Heart Mountain slide, Wyoming, USA. The Journal of Geology, 125:439-457.

Malone, D.H., Craddock, J.P., Wallenberg, A., Gaschot, B., and Luczaj, J.A. 2022. Geology of Chief Joseph Pass, Wyoming: Crest of Rattlesnake Mountain anticline and escape path of the Eocene Heart Mountain slide; pp. 313-333 in Craddock, J.P., Malone, D.H., Foreman, B.Z., and Konstantinou, A. (eds.), *Tectonic Evolution of the Sevier-Laramide Hinterland, Thrust Belt, and Foreland, and Postorogenic Slab Rollback (180-20 Ma)*. Geological Society of America Special Paper 555.

Malone, D.H., Schroeder, K., and Craddock, J.P. 2014. Detrital zircon age and provenance of Wapiti Formation tuffaceous sandstones, South Fork Shoshone River Valley, Wyoming. The Mountain Geologist, 51:271-286.

Malone, D.H., and Sundell, K.A. 2000. Gigantic Eocene landslide/debris-avalanche deposits of Wyoming's eastern Absaroka Range. Geological Society of America Abstracts with Programs, 32:A119.

Malone, J.R., Malone, D.H., and Craddock, J.P. 2022. Sediment provenance and stratigraphic correlations of the Paleogene White River Group in the Bighorn Mountains, Wyoming. The Mountain Geologist, 59:273-293.

Max Planck Institute for the Science of Human History. 2020. Human populations survived the Toba volcanic supereruption 74,000 years ago. Science Daily, February 25, 2020.

Mathison, M.E., Hummer, D., Bown, T.M., and Warner, A.J. 2025. Manuscript. Some physicochemical properties of breccias formed by movement on the Shoshone/Sunlight/Abiathar Detachment Fault and The Heart Mountain/McCullough Peaks Sturzstrom, northwest Wyoming, USA.

McGee, W.J. 1897. Sheetflood erosion. Geological Society of America Bulletin, 8:87-112.

McKenna, M.C. 1980. Remaining evidence of Oligocene sedimentary rocks previously present across in Bighorn Basin, Wyoming; pp. 143-145 in Gingerich, P.D. (ed.), *Early Cenozoic Paleontology and Stratigraphy of the Bighorn Basin, Wyoming*. University of Michigan Museum Papers on Paleontology, no. 24.

McKenna, M.C., and Love, J.D. 1972. High-level strata containing early Miocene mammals on the Bighorn Mountains, Wyoming. American Museum Novitates, no. 2490:1-31.

Mears, Jr., B. 1993. Geomorphic history of Wyoming and high-level erosion surfaces; pp. 608-626 in Snoke, A.W., Steidtmann, J.R., and Roberts, S.M. (eds.) *Geology of Wyoming* (Blackstone/Love Volume), Geological Survey of Wyoming Memoir 5.

Melosh, H.J. 1983. Acoustic fluidization: can sound waves explain why rock debris appears to flow like a fluid in some energetic geologic events? American Scientist, 71:158-165.

Melosh, H.J. 2018. Acoustic fluidization. American Scientist, 71:158-165.

Merrill, R.D. 1974. Geomorphology of terrace remnants of the Greybull River, Big Horn Basin, northwestern Wyoming. Austin, TX, University of Texas, PhD Thesis: 267 pp.

Mitchell, T.M., Smith, S.A.F., Anders, M.H., Di Toro, G., Nielsen, S., Cavallo, A., and Beard, A.D. 2015. Catastrophic emplacement of giant landslides aided by thermal decomposition; Heart Mountain, Wyoming. Earth and Planetary Science Letters, 411:199-207.

Morrill, B.J. 1971. Evidence of record vertical accelerations at Kagel Canyon during the earthquake; pp. 177-181 in The San Fernando, California, Earthquake of February 9, 1971, U.S. Geological Survey Professional Paper 733.

Morris, J. 2013. Mount Toba, the eruption. PREZI, Tuesday, 12/3/2013.

Moss, J.H., and Bonini, W.E. 1961. Seismic evidence supporting a new interpretation of the Cody terrace near Cody, Wyoming. Geological Society of America Bulletin, 72:547-555.

Moye, F.J., Hackett, W.R., Blakley, J.D., and Snider, L.G. 1988. Regional geologic setting and volcanic stratigraphy of the Challis volcanic field, central Idaho; pp. 87-97 *in* Winkler, G.R. and Hackett, W.R. (eds.), Guidebook to the Geology of Central and Southern Idaho. Idaho Geological Survey Bulletin, 27.

Nelson, P.L., and Grand, S.P. 2018. Lower-mantle plume beneath the Yellowstone hot spot revealed by core waves. Nature Geoscience, 11:280-284.

Nelson, W.H. 1991. Kinematic model of a continuous Heart Mountain allochthon: Discussion and reply. Geological Society of America Bulletin, 103:718-722.

Nelson, W.H., and Pierce, W.G. 1968. Wapiti Formation and Trout Peak Trachyandesite, northwest Wyoming. U.S. Geological Survey Bulletin, 1254-H:H1-H11.

Nelson, W.H., Pierce, W.G., Parsons, W.H., and Brophy, G.P. 1972. Igneous activity, metamorphism, and Heart Mountain faulting at White Mountain, northwestern Wyoming: Geological Society of America Bulletin, 83:2607-2620.

Oberholzer, J. 1933. Geologie der Glarneralpen. Beitr. Z. Geol. Karte d. Schweiz. N.F. Leif., 28.

O'Neill, W.A. 1980. $^{40}Ar^{39}Ar$ ages of selected tuffs of the Green River Formation, Wyoming, Colorado, and Utah. M.S. Thesis, The Ohio State University:1-152.

Oppenheimer, C. 2002. Limited global change due to the largest known Quaternary eruption, Toba ~74 Kyr BP? Quaternary Science Reviews, 21:1593-1609.

Oppenheimer, C. 2003. Climatic, environmental, and human consequences of the largest known historic eruption: Tambora Volcano (Indonesia), 1815. Progress in Physical Geography, 27:230-259.

Palmquist, R.C. 1978. Ash dated terrace sequence in the eastern portion of the Bighorn Basin, Wyoming. Geological Society of America Abstracts with Programs, 10:467.

Palmquist, R.C. 1983. Terrace chronologies in the Bighorn Basin, Wyoming. Wyoming Geological Association, 34th Annual Field Conference Guidebook:217-231.

Philip, H., and Ritz, J.-F. 1999. Gigantic paleolandslide associated with active faulting along the Bogd Fault (Gobi-Altay, Mongolia). Geology, 27:211-214.

Pierce, W.G. 1941. The Heart Mountain and South Fork thrusts, Park County, Wyoming. American Association of Petroleum Geologists Bulletin, 25:2021-2045.

Pierce, W.G. 1957. Heart Mountain and South Fork detachment thrusts of Wyoming. American Association of Petroleum Geologists Bulletin, 41:591-626.

Pierce, W.G. 1960. The "break-away" point of the Heart Mountain detachment fault in northwestern Wyoming. U.S. Geological Survey Professional Paper 400-B:B236-B237.

Pierce, W.G. 1963a. Cathedral Cliffs Formation, the Early Acid Breccia Unit of northwestern Wyoming. Geological Society of America Bulletin, 74:9-22.

Pierce, W.G. 1963b. Reef Creek detachment fault, northwestern Wyoming. Geological Society of America Bulletin, 74:1225-1236.

Pierce, W.G. 1965a. Geologic map of the Clark Quadrangle, Park County, Wyoming. U.S. Geological Survey Geologic Quadrangle Map, GQ-477.

Pierce, W.G. 1965b. Geologic map of the Deep Lake Quadrangle, Park County, Wyoming. U.S. Geological Survey Geologic Quadrangle Map, GQ-478.

Pierce, W.G. 1966a. Geologic map of the Cody Quadrangle, Park County, Wyoming. U.S. Geological Survey Geologic Quadrangle Map, GQ-542.

Pierce, W.G. 1966b. Role of fluid pressure in mechanics of overthrust faulting: Discussion. Geological Society of America Bulletin, 77:565-568.

Pierce, W.G. 1968a. Tecronic denudation as exemplified by the Heart Mountan fault, Wyoming; pp. 191-197 in Orogenic Belts. Proceedings of the 23rd International Geological Congress.

Pierce, W.G. 1968b. The Carter Mountain landslide area, northwest Wyoming; pp. D235-D241 in Geological Survey Research, 1968; U.S. Geological Survey Professional Paper 600-D.

Pierce, W.G. 1973. Principal features of the Heart Mountain fault, and the mechanism problem; pp. 457-471 in DeJong, K.A. and Scholten, R. (eds.) *Gravity and Tectonics*. New York, Wiley.

Pierce, W.G. 1978. Geologic map of the Cody 1° X 2° Quadrangle, northwestern Wyoming. U.S. Geological Survey Miscellaneous Field Studies Map, MF-963.

Pierce, W.G. 1979. Clastic dikes of Heart Mountain fault breccia, northwestern Wyoming, and their significance. U.S. Geological Survey Professional Paper 1133:1-25.

Pierce, W.G. 1980. The Heart Mountain break-away fault, northwestern Wyoming. Geological Society of America Bulletin, 91:272-281.

Pierce, W.G. 1982. Relation of volcanic rocks to the Heart Mountain fault. Wyoming Geological Association 33rd Annual Field Conference Guidebook: 181-184.

Pierce, W.G. 1985. Map showing present configuration of the Heart Mountain fault surface and related features, Wyoming and Montana. Geological Survey of Wyoming Map Series, no. 15.

Pierce, W.G. 1987. The case for tectonic denudation by the Heart Mountain fault—a response. Geological Society of America Bulletin, 99:552-568.

Pierce, W.G. 1997. Geologic map of the Cody 1 x 2° Quadrangle, northwestern Wyoming. U.S. Geological Survey Miscellaneous Investigations Map, I-2500.

Pierce, W.G., and Andrews, D.A. 1941. Geology and oil and coal resources of the region south of Cody, Park County, Wyoming. U.S. Geological Survey Bulletin, 921-B:99-180.

Pierce, W.G., and Nelson, W.H. 1968. Geologic Map of the Pat O'Hara Mountain Quadrangle, Park County, Wyoming. U.S. Geological Survey Geologic Quadrangle Map, GQ-755.

Pierce, W.G., and Nelson, W.H. 1969. Geologic map of the Wapiti Quadrangle, Park County, Wyoming. U.S. Geological Survey Geologic Quadrangle map, GQ-778.

Pierce, W.G., and Nelson, W.H. 1973. Crandall Conglomerate, an unusual stream deposit, and its relation to Heart Mountain faulting. Geological Society of America Bulletin, 84:2631-2644.

Pierce, W.G., and Nelson, W.H. 1986. Some features indicating tectonic denudation by the Heart Mountain fault. Montana Geological Survey/Yellowstone Bighorn Research Association Field Conference Guidebook:155-164.

Pierce, W.G., Nelson, W.H., and Prostka, H.J. 1973. Geologic map of the Pilot Peak Quadrangle, Park County, Wyoming. U.S. Geological Survey Miscellaneous Geologic Investigations Map, I-816.

Pierce, W.G., Nelson, W.H., and Prostka, H.J. 1982. Geologic map of the Dead Indian Peak Quadrangle, Park County, Wyoming. U.S. Geological Survey Geological Quadrangle Map, GQ-1564.

Pollet, N., Cojean, R., Couture, R., Schneider, J.-L., Strom, A.L., Voirin, C., and Wassmer, P. 2005. A slab-on-slab model for the Flims rockslide (Swiss Alps). Canadian Geotechnical Journal, 42:587-600.

Pollet, N., and Schneider, J.-L.M. 2004. Dynamic disintegration processes accompanying transport of the Holocene Flims sturzstrom (Swiss Alps). Earth and Planetary Science Letters, 221:433-448.

Prostka, H.J. 1978. Heart Mountain fault and Absaroka volcanism, Wyoming and Montana, U.S.A.; pp. 423-437 in Voight, B. (ed.), Rockslides and Avalanches, I. Natural Phenomena. New York, Elsevier.

Prostka, H.J., Ruppel, E.T., and Christiansen, R.L. 1975. Geologic map of the Abiathar Peak Quadrangle, Yellowstone National Park, Wyoming and Montana. U.S. Geological Survey Geologic Quadrangle Map, GQ-1244.

Radbruch-Hall, D.H. 1978. Gravitational creep of rock masses on slopes; pp. 607-657 in B. Voight (ed.), Rockslides and Avalanches, I. Natural Phenomena. *Developments in Geotechnical Engineering*, vol. 14A. New York, Elsevier.

Rampino, M.R., and Ambrose, S.H. 2000. Volcanic winter in the Garden of Eden: The Toba supereruption and the late Pleistocene human population crash. Geological Society of America Special Paper 345:71-82.

Reheis, M.C. 1983. Glaciofluvial origin and drainage history revealed by terraces in the northern Bighorn Basin, Montana. Geological Society of America Abstracts with Programs, 15 (5):431.

Reheis, M.C. 1985. Evidence for Quaternary tectonism in the northern Bighorn Basin, Wyoming and Montana. Geology, 13:364-367.

Rempel, A.W., and Rice, J.R. 2006. Thermal pressurization and onset of melting in fault zones. Journal of Geophysical Research, 111:18 pp. B09314, doi:10.1029/2006JB004314.

Rengers, N. 1970. Influence of surface roughness on the friction properties of rock planes. Proceedings of the Second International Congress of the Society of Rock Mechanics, 1:229-234.

Rhodes, M.K., Malone, D.H., Carroll, A.R., and Smith, E.M. 2007. Sudden desiccation of Lake Gosiute at ~49 Ma: A downstream record of Heart Mountain faulting? The Mountain Geologist, 44:1-10.

Ritter, D.F. 1967. Terrace development along the front of the Beartooth Mountains, southern Montana. Geological Society of America Bulletin, 78:467-484.

Ritter, D.F. 1975. New information concerning the geomorphic evolution of the Bighorn Basin. Wyoming Geological Association 27th Annual Field Conference Guidebook:37-44.

Roberts, N.J., and Evans, S.G. 2013. The gigantic Saymareh (Saidmarreh) rock avalanche, Zagros Fold-Thrust Belt, Iran. Journal of the Geological Society of London, 170:685-700.

Robinove, C.J., and Langford, R.H. 1963. Geology and ground water resources of the Greybull River—Dry Creek area, Wyoming. U.S. Geological Survey Water Supply Paper, no. 1596.

Rohrer, W.L. 1966. Geology of the Adam Weiss Peak Quadrangle, Hot Springs and Park Counties, Wyoming. U.S. Geological Survey Bulletin, 1241-A:1-39.

Rohrer, W.L., and Leopold, E.B. 1963. Fenton Pass Formation (Pleistocene?), Bighorn Basin, Wyoming. U.S. Geological Survey Professional Paper 475-C:C45-C48.

Rohrer, W.L., and Smith, J.W. 1969. Tatman Formation. Wyoming Geological Association 21st Annual Field Conference Guidebook:49-54.

Rouse, J.T. 1937. Genesis and structural relationships of the Absaroka volcanic rocks, Wyoming. Geological Society of America Bulletin, 48:1257-1296.

Rouse, J.T. 1940. Structural and volcanic problems in the southern Absaroka Mountains, Wyoming. Geological Society of America Bulletin, 51:1413-1428.

Sales, J.K. 1983. Heart Mountain—blocks in a giant volcanic rock glacier. Wyoming Geological Association 34th Annual Field Conference Guidebook, pp. 117-165.

Sataki, K., and Atwater, B.F. 2007. Long term perspectives on giant earthquakes and tsunamis at subduction zones. Annual Review of Earth and Planetary Sciences, 35:349-374.

Schneider, J.-L., Wassmer, P., and Lédesert, B. 1999. La fabrique interne des dépôts du sturzstrom fe Flims (Alpes suisses) caratéristiques et impications sur les mécanismes de transportation. Comptes rendus del'Académie des sciences, Paris, Earth and Planetary Sciences, 328:607-613.

Schumm, S.A. 1967. Rates of surficial rock creep on hillslopes in western Colorado. Science, 155:560-561.

Seeland, D.A. 1978. Eocene fluvial drainage patterns and their implications for uranium and hydrocarbon exploration in the Wind River Basin, Wyoming. U.S. Geological Survey Bulletin, 1446:1-21.

Seeland, D. 1998. Late Cretaceous, Paleocene, and early Eocene paleogeography of the Bighorn Basin and northwest Wyoming. Wyoming Geological Association 49th Annual Field Conference Guidebook:137-165.

Shaller, P.J., and Smith-Shaller, A.S. 1996. Review of proposed mechanisms for sturzstroms (long-runout landslides); pp. 185-202 in P.L. Abbott and D.C. Seymour (eds.), Sturzstroms and Detachment Faults, Anza-Borrego Desert State Park, California. Santa Ana California, South Coast Geological Society, Inc.

Shoaei, Z., and Ghayoumian, J. 1998. Seimareh landslide, the largest complex slide in the world; pp. 1337-1342 in Moore, D. and O. Hungr (eds.), Proceedings of the Eighth International Congress of the International Association for Engineering Geology and the Environment.

Shreve, R. 1968. The Blackhawk Landslide. Geological Society of America Special Paper, 108:1-47.

Siebert, L. 1984. Large volcanic debris-avalanches: characteristics of source areas, deposits, and associated eruptions. Journal of Volcanology and Geothermal Research, 22:163-197.

Siebert, L., Glicken, H., and Ui, T. 1987. Volcanic hazards from Bezymianny- and Bandai-type eruptions. Volcanology Bulletin, 49:435-459.

Sinclair, W.J., and Granger, W. 1912. Notes on the Tertiary deposits of the Bighorn Basin. American Museum of Natural History Bulletin, 31:57-67.

Smedes, H.W., and Prostka, H.J. 1972. Stratigraphic framework of the Absaroka Volcanic Supergroup in the Yellowstone National Park region. U.S. Geological Survey Professional Paper, 729-C:C1-C33.

Smith, M.E., Carroll, A.R., Jicha, B.R., Cassel, E.L., and Scott, J.J. 2014. Paleogeographic record of Eocene Farallon slab rollback beneath western North America. Geology, 42:1039-1042.

Solonenko, V.P. 1977. Landslides and collapses in seismic zones and their prediction. International Association of Engineering Geologists Bulletin, 15:4-8.

Steinberger, B., Nelson, P.L., Grand, S.P., and Wang, W. 2019. Yellowstone plume conduit tilt caused by large-scale mantle flow. Geochemistry, Geophysics, Geosystems, 20:5896-5912.

Stevens, E.H. 1938. Geology of the Sheep Mountain remnant of the Heart Mountain thrust sheet, Park County, Wyoming. Geological Society of America Bulletin, 49:1233-1266.

Stothers, R.B. 1984. The great Tambora eruption in 1815 and its aftermath. Science, 224:1191-1198.

Stover, C.W., and Coffman, J.L. 1993. Seismicity of the United States, 1568-1989 (revised). U.S. Geological Survey Professional Paper, 1527:1-418.

Straw, W.T., and Schmidt, C.J. 1981. Heart Mountain detachment fault: A phreatomagmatic-hydraulic hypothesis? Geological Society of America Abstracts with Programs, 13:562.

Sundell, K.A. 1990. Sedimentation and tectonics of the Absaroka Basin of northwestern Wyoming. Wyoming Geological Association 41st Annual Field Conference Guidebook:105-122.

Sundell, K.A. 1993. A geologic overview of the Absaroka volcanic province; pp. 480-506 in Snoke, A.W., Steidtmann, J.R., and Roberts, S.M. (eds.), Geology of Wyoming. Geological Survey of Wyoming Memoir no. 5.

Surdam, R.C., and Stanley, K.O. 1980. Effect of changes in drainage-basin boundaries on sedimentation in Eocene lakes Gosiute and Uinta of Wyoming, Utah, and Colorado. Geology, 8:135-139.

Swallow, E.J., Wilson, C.J.N., Charlier, B.I.A., and Gamble, J.A. 2019. The Huckleberry Ridge Tuff, Yellowstone: evacuation of multiple magmatic systems in a complex episodic eruption. Journal of Petrology, 60:1371-1426.

Swanson, M. E., Wernicke, B.P., and Hauge, T.A. 2016. Episodic dissolution, precipitation, and slip along the Heart Mountain detachment, Wyoming. Journal of Geology, 124:75-97.

Templeton, A.S., Sweeney, J. Jr., Manske, H., Tilghman, J.F., Calhoun, S.C., Violich, A., and Chamberlain, C.P. 1995. Fluids and the Heart Mountain fault revisited. Geology, 23:929-932.

Terzaghi, K. 1950. Mechanism of landslides. Geological Society of America Engineering Geology (Berkey) Volume:83-123.

Tourtelot, H.A. 1946. Tertiary stratigraphy in the northeastern part of the Wind River Basin, Wyoming. U.S. Geological Survey Oil and Gas Investigations Preliminary Chart 22.

Twidale, C.R. 1981. Origins and environments of pediments. Journal of the Geological Society of Australia, 28:423-434.

Twidale, C.R. 2014. Pediments and platforms: problems and solutions. Géomorphologie: relief, processus, environment, 20:43-56.

Ui, T. 1985. Debris avalanche deposits associated with volcanic activity; pp. 405-410 in Proceedings of the 4th International Conference and Field Workshop on Landslides. Tokyo, The Japanese Landslide Society.

Varnes, D.J. 1978. Slope movement types and processes; pp. 11-33 in Schuster, R.L., and Krizek, R.J. (eds.), Landslides: Analysis and Control. National Resources Council Transportation Research Board Special Report 176.

Voight, B. 1972. Fluid-wedge hypothesis and the Heart Mountain and Reef Creek decollements, northwestern Wyoming, USA. Geological Society of America Abstracts with Programs, 4:698.

Voight, B. 1973a. Clastic fluidization phenomena and the role of fluid pressure in mechanics of natural rock deformation. Geological Society of America Northeastern Section Meeting Abstracts with Programs 5:233.

Voight, B. 1973b. Role of fluid pressure in mechanics of South Fork, Reef Creek, and Heart Mountain rockslides. Geological Society of America, Northeastern Section, Abstracts with Programs, 5:233-234.

Voight, B. 1973c. The mechanics of retrogressive block-gliding, with emphasis of the evolution of the Turnagain Heights landslide, Anchorage, Alaska; pp. 97-121 in de Jong, K.A. and R. Scholten (eds.), Gravity and Tectonics. New York, Wiley-Interscience; 502 pp.

Voight, B. 1974. Architecture and mechanics of the Heart Mountain and South Fork rockslides; pp. 26-36 in B. Voight and M.A. Voight (eds.), Rock Mechanics, the American Northwest. University Park, PA, Pennsylvania State University.

Voight, B. 1978. Chapter 3 – Lower Gros Ventre Slide, Wyoming, USA; pp. 113-162 in Voight, B. (ed.), Rockslides and Avalanches, Developments in Geotechnical Engineering, vol. 14.

Wassmer, P., Schneider, J.-L., and Pollet, N. 2002. Internal structure of huge mass movements: a key for a better understanding on long runout. The multi-slab theoretical model. UNESCO International Symposium on Landslide Risk Mitigation and Protection of Cultural and Natural Heritage, Kyoto, Japan. Tokyo, The Japan Landslide Society, pp. 97-107.

Weidinger, T.W., Schramm, J.-M., and Surenian, R. 1996. On preparatory causal factors, initiating the prehistoric Tsergo Ri landslide (Langthang Himal, Nepal). Tectonophysics, 260:95-107.

Welch, J.L., Foreman, B.Z., Malone, D.H., and Craddock, J.P. 2022. Provenance of early Paleogene strata in the Bighorn Basin (Wyoming, USA): Implications for Laramide tectonism and basin-scale stratigraphic patterns; pp. 241-264 in Craddock, J.P., Malone, D.H., Foreman, B.Z., and Konstantinou, A. (eds.), Tectonic Evolution of the Sevier-Laramide Hinterland, Thrust Belt, and Foreland, and Postorogenic Slab Rollback (180-20 Ma). Geological Society of America Special Paper 555.

Westerhold, T., Röhl, U., Wilkens, R.H., Gingerich, P.D., Clyde, W.C., Wing, S.L., Bowen, G.J., and Kraus, M.J. 2018. Synchronizing early Eocene deep-sea and continental records—cyclostratigraphic age models for the Bighorn Basin Coring Project drill cores. Climate of the Past, 14:303-319.

Wilson, W.H. 1964. Geologic reconnaissance of the southern Absaroka Mountains, northwest Wyoming: Part I—The Wood River-Greybull River area. Contributions to Geology (The University of Wyoming), 3:60-77.

Wilson, W.H. 1970. Geologic Map of the Soapy Dale Peak Quadrangle, Hot Springs County, Wyoming. Wyoming Geological Survey Map, 1:24000.

Wilson, W.H. 1971. Volcanic geology and mineralization, Absaroka Mountains, northwest Wyoming. Wyoming Geological Association, 23rd Annual Field Conference Guidebook:151-155.

Wilson, W.H. 1975a. Detachment faulting in volcanic rocks, Wood River area, Park County, Wyoming. Wyoming Geological Association 27th Annual Field Conference Guidebook:167-171.

Wilson, W.H. 1975b. The copper-bearing Meadow Creek Granodiorite, upper Wood River area, Park County, Wyoming. Wyoming Geological Association, 27th Annual Field Conference Guidebook:235-241.

Wilson, W.H. 1982. Geologic Map of the Dick Creek Lakes, Dunrud Peak, Francs Peak, Noon Point, and Twin Peaks Quadrangles, Fremont, Hot Springs, and Park Counties, Wyoming. Wyoming Geological Survey, Map Series, no. 10.

Wing, S.L., Bown, T.M., and Obradovich, J.D. 1991. Early Eocene biotic and climatic change in interior western North America. Geology, 19:1189-1192.

Yost, C.L., Jackson, L.J., Stone, J.R., and Cohen, A.S. 2018. Subdecadal phytolith and charcoal records from Lake Malawi, East Africa imply minimal effects on human evolution from the ~74 Ka supereruption. Journal of Human Evolution, 116:75-94.

Yumul, G.P., Dimalanta, C.B., Maglambayam, V.B., and Marquez E.J. 2008. Tectonic setting of a composite terrane: A review of the Philippine island arc system. Geosciences Journal, 12:7-17.

Welch, J.L. 2021. Provenance of early Paleogene strata in the Bighorn Basin (Wyoming, U.S.A.): Implications for Laramide tectonism and basin-scale stratigraphic patterns. Western Washington University Master of Science Thesis, 65 pp.

Wolfe, J.A. 1977. Large, Holocene low-angle landslide, Samar Island, Philippines. Geological Society of America, Reviews in Engineering Geology, III:149-153.

Zhou, Q., Liu, L., and Hu, J. 2018. Western US volcanism due to intruding oceanic mantle driven by ancient Farallon slabs. Nature Geoscience, 11:70-76.

TABLE I

Important dates of phenomena related to the Shoshone/Sunshine/Abathair Detachment Fault (SSADF) and the Heart Mountain/McCullough Peaks Sturzstrom (HMMPS).

~2.08 Ma—Average age of several episodes of emplacement of Huckleberry Ridge Tuff deposition (Swallow *et al.*, 2019) and probable time of emplacement of ECOCDA (Bown and Love, 1987) and HMMPS deposit (this paper).

2.58 Ma—Age of Pliocene/Pleistocene boundary (Gibbard and Head, 2009).

43.95 Ma—Age of youngest Eocene volcanic activity in the Absaroka Range (Hiza, 1999a).

48.5 +/- 1.2 Ma – Age of some allochthonous volcanic rocks of Aycross Formation on Squaw Buttes (Malone, Craddock, and Mathesin, 2014).

48.87 +/- 0.2 Ma—Age of displacement on HMDF (Shoshone/Sunlight/Abiathar Detachment Fault) according to Malone, Craddock, Anders, and Wulff (2014).

48.99 Ma—Age of upper part of lower Wapiti Formation (Malone, Schroeder, and Craddock, 2014).

49.19 Ma –Age of displacement on HMDF (=SSADF) according to Malone *et al.* (2022)—see Malone *et al.* (2017) for age estimate of 48.97-49.19 Ma.

49.44 Ma—Age of middle part of lower Wapiti Formation (Malone, Schroeder, and Craddock, 2014).

~49.5 Ma—Estimated age of displacement of Shoshone/Sunlight/Abiathar Detachment Fault (this paper; = younger than displaced Crandall Conglomerate and older than the base of the overlying Wapiti Formation).

49.57 (+0.51/-0.64) Ma—Age of Crandall Conglomerate (Malone, Breedan, Craddock, Anders, and MacNamee, 2015).

50.05 +/- 0.65 Ma—Age of Aycross Formation based on zircon geochronology from sample taken near Togwotee Pass (Malone, Craddock, Garber, and Trela (2017).

50.4 +/- 0.5 Ma—Age of part of type Aycross Formation (Love, McKenna, and Dawson, 1976; revised from date of 49.2 +/- 0.5 Ma by Obradovich *in* Bown, 1982a:A19).

50.53 (+1.59/-0.78) Ma—Age of some allochthonous rocks on Squaw Buttes (Malone, Craddock, and Mathesin, 2014).

50.84 (+1.29/-0.78) Ma—Age of upper part of Tatman Formation beneath allochthonous mass on Squaw Buttes (Malone, Craddock, and Mathesin, 2014).

52.1 +/- 1.8 Ma—Oldest zircon date for Tatman Formation (Malone, Craddock, and Mathesin, 2014).

52.8 +/- 0.3 Ma—Age of tuff at 634-meter level of 770 meter-thick Willwood Formation (Wing, Bown, and Obradovich, 1991; Bown *et al.*, 1994).

TABLE II

Locations of principal surviving remnants of Heart Mountain/McCullough Peaks Sturzstrom debris situated between the Shoshone River and the top of Middle Peak of McCullough Peaks (Corbett Masses = CDM #1-11), showing estimated maximum preserved thicknesses of debris and some approximate dimensions of deposit remnants.

CDM #1: Main Corbett mass (~18 m thick, ~1,925 m long); east end = 44° 36' 56.41" N, 108° 54' 44.14" W; west end = 44° 37' 08.95" N, 108° 56' 01.32" W.

CDM #2: (35-40 m thick; ~500 m long); east end = 44° 37' 24.95" N, 108° 55' 48.11" W; west end = 44° 37' 32.12" N, 108° 56' 13.83" W.

CDM #3: (~13 m thick), 44° 37' 33.24" N, 108° 56' 10.16" W.

CDM #4: (~4 m thick), 44° 37' 24.58" N, 108° 55' 44.15" W.

CDM #5: (8 m thick), 44° 37' 18.59" N, 108° 55' 39.87" W.

CDM #6: (9 m thick; ~260 m long); east end = 44° 37' 14.52" N, 108° 55' 53.91"W; west end = 44° 37' 13.42" N, 108° 56' 06.28" W.

CDM #7: (~3 m thick), 44° 37' 10.15" N, 108° 55' 44.64" W.

CDM #8: (~6 m thick), 44° 37' 01.52" N, 108° 55' 33.63" W.

CDM #9: (~16 m thick), 44° 36' 59.39" N, 108° 54' 46.06" W.

CDM #10: (~3 m thick), 44° 37' 06.75" N, 108° 55' 8.60" W.

CDM #11: (~4 m thick), 44° 36' 53.55" N, 108° 54' 48.63" W.

TABLE III

Estimated ages of terraces and other surfaces in the western Bighorn Basin, calculated using estimated erosion rates determined by Palmquist (1978 = **P**) and Love and Walsh (1981 = **L&W**), respectively.

1) **Polecat Bench terrace, at 5205 feet/1587 m):** The top of the south side of the bench at 44° 46' 24" N, 108° 52' 45.0" W is 262 m above a closely adjacent part of the bed of the Shoshone River at 44° 42' 15" N, 108° 49' 11.23" W. Estimated age of terrace = approximately **1,430,000 years (P, see also Reheis, 1985 at 1,400,000 years), and 1,288,525 years (L&W)**. (Geologic information supplied by Reheis [1983] suggests a somewhat younger age of the Polecat Bench surface).

2) **Corbett Masses (top of largest mass, at 5986 feet/1825 m):** at 44° 36' 53.91" N, 108° 54' 48.74" W is 425 m above the bed of the adjacent Shoshone River at 44° 37' 01.54" N, 108° 56' 51.47" W. Estimated age of emplacement of mass (and of HMMPS) = approximately **2,245,000 years (P), and 2,091,000 years (L&W)**.

3) **Y-U Bench terrace, at 5688 feet/1734 m:** at 44° 18' 41.59" N, 108° 45' 51.95" W lies 161 m above the bed of the adjacent Greybull River at 44° 18' 08.27" N, 108° 44' 13.21" W. Estimated age of terrace = approximately **925,000 years (P), and 791,803 years (L&W)**.

4) **Tatman Mountain surface (top of north side, at 6268 feet/1911 m):** at 44° 17' 12.88" N, 108° 35' 05.71" W is 399 m above the bed of the adjacent Greybull River at 44° 20' 45.63" N, 108° 40' 11.05" W. Estimated age of surface = approximately **2,115,000 years (P), and 1,962,295 years (L&W)**. Based on fossil pollen, Rohrer and Leopold (1963) determined the age of the Fenton Pass Formation, a pebble and cobble terrace deposit that forms the Tatman surface, to be of Pliocene or Pleistocene age; however, studies of late Pleistocene terraces by Moss and Bonini (1961) indicate a physiographic age of no older than early Pleistocene.

5) **Squaw Buttes (top of west Squaw Butte, at 6,126 feet/1868 m):** at 44° 06' 33.90" N, 108° 34' 35.44" W lies 364 m above the bed of the adjacent Fifteenmile Creek (though intermittent, the closest major stream) at 44° 11' 56.68" N, 108° 34' 53.02" W. (The Wood River once probably flowed east across the Bighorn Basin—see note below). Estimated age of surface = approximately **2,065,122 years (P), and 1,913,235 years (L&W)**.

6) **Red Butte (top of butte, at 5,151 feet/1690 m):** at 44° 12' 57.71" N, 108° 14' 33.31" W lies 279 m above the bed of the Greybull River (the closest major stream) at 44° 23' 24" N, 108° 20' 16" W. Estimated age of terrace deposit = approximately **1,515,000 years (P), and 1,372,131 (L&W)**.

7) **Carter Mountain Exotic Mass (top of mass, at 6,888 feet/2100 m):** at 44° 21' 19.48" N, 109° 10' 45.18" W lies 446 m above the bed of the South Fork of the Shoshone River (the closest major stream) at 44° 25' 55.22" N, 109° 15' 13.53" W. Estimated age of emplacement = approximately **2,350,000 years (P), and 2,194,000 years (L&W).**

8) **Elk Butte terrace (highest point, at 6,210 feet/1893 m):** at 44° 21' 09.29" N, 108° 55' 35.83" W lies 276 m above the bed of the adjacent Greybull River (the closest major stream) at 44° 16' 10.82" N, 108° 46' 57.8" W. Estimated age of terrace = approximately **1,500,000 years (P), and 1,356,000 years (L & W).**

TABLE IV

Estimates of percent lithologic compositions of gravels and other clasts on surfaces at several localities bordering Dry Creek and at one locality south of Heart Mountain. **V** = volcanic rocks; **Q** = quartzite; **S** = sandstone; **C** = chert; **PZ** = Paleozoic micrite and/or dolomite. Estimates are based on examination of clasts in the 5 cm through cobble size ranges on approximately 20-minute walkabouts by two researchers. Note: Where they occur, volcanic clasts are much more abundant than shown in the smaller clast sizes. Numbers refer to localities in field notes; such that Stop #10-19 = the tenth stop of field year 2019 and Stop 2-21 = the second stop of field year 2021). Note: percentages are rough estimates and composition figures may not add up to 100%.

PEDIMENT SURFACE ON SOUTH SIDE OF HEART MOUNTAIN

Stop #90-21 (elev = 5678'): 44° 36' 27.48" N, 109° 07' 7.35" W
    Q = 12%; C = 80%; PZ = 8% (largest of all clasts—up to 92 cm diameter)

SWALE CONNECTING SAGE CREEK WITH THE NORTH FORK OF DRY CREEK

Stop #10-19 (elev = 5140'): 44° 27' 57" N, 108° 51' 00" W
    Swale littered with 100% angular and subangular PZ cobble and pebble clasts

Stop #88-21 (elev = 5286'): 44° 28' 6.66" N, 108° 52' 10.38" W
    V = 75%; Q = 10%; S and C = 2-3%; PZ = 7%

NORTH FORK DRY CREEK; stream terraces

Stop #2-21 (elev = 5443'): 44° 22' 27.41" N, 108° 52' 22.04" W
    V= 75%; Q = 10%; C = 2%; PZ = 8%; (24 x 18 cm ; s-ang.)

Stop #11-19 (elev = 5110'): 44° 25' 38.2" N, 108° 49' 35.9" W
    PZ boulders in channel of N. Fork Dry Creek (diameter of largest = 81 cm)

Stop #84-21 (elev = 5367'): 44°24' 0.75" N, 108° 53' 3.19" W
    V = 65%; Q = 20%; C = 1%; PZ = 8%

Stop #85-21 (elev = 5349'): 44° 24' 6.2" N, 108° 52' 28.77" W
    V = 65%; Q = 18%; C = 1%; PZ = 8%
    Locally derived Cretaceous sandstone = 2%

Stop #86-21 (elev = 5327'): 44° 24' 5.43" N, 108° 52' 22.32" W
    V = 70%; Q = 20%; PZ = 8%

Stop #87-21 (elev = 5308'): 44° 24' 5.85" N, 108° 52' 18.03" W
    V = 80%; Q = 10%; PZ = 3%
    Rusty ss concretions derived from Cretaceous rocks = 4-5%

Stop #1-22, Elk Butte SW (elev = 6203'): 44° 21' 08.7" N, 108° 55' 31.2" W
    V = 60+%; PZ = 35%

Stop #1a-22, Elk Butte east (elev = 6167'): 44° 21' 11.7" N, 108° 55' 18.4" W
    V = 80%; PZ = 15+%

Stop #13-22 Lowest terrace on North Fork Dry Creek (elev = 5235'): 44° 24' 58.4" N, 108° 51' 54.8" W
    V = 65%; Q = 15%; PZ = 15%; C = <1%; S = ~5% (locally derived)

SOUTH FORK DRY CREEK; stream terraces

Stop #27-21 (elev = 5661'): 44° 19' 38.86" N, 108° 55' 39.83" W
    V = 80%; Q = 3-4%; C = trace; PZ = 8%
    Locally derived Mesaverde sandstone = 5-7%

Stop #28-21 (elev = 5496'): 44° 19' 6.47 " N, 108° 53' 39.65" W
    V = 45%; Q = 45%; PZ = 1%
    Locally derived Mesaverde sandstone = 5%

Stop #29-21 (elev = 5712'): 44° 21' 11.91" N, 108° 54' 14.8" W
    V = 75%; Q = 15%; PZ = 7%
    Locally derived Mesaverde sandstone = 3%

DRY CREEK (east of confluence of North and South forks); stream terraces

Stop #14-19 (elev = 4177'): 44° 31' 28.3" N, 108°09' 15" W
    V = >95%; Q = 5%; PZ = 0%

Stop #15-19 (elev = 4176'): 44°31' 31.2" N, 108° 09' 0.1" W
    V = >95%; Q = ~1%; PZ = 0%

Stop #16-19 (elev = 4221'): 44° 30' 32.4" N, 108° 14' 08.5" W
    V = 80%; Q = 15%; PZ = 0%

Stop #18-19 (elev = 4248'): 44° 31' 04" N, 108° 15' 0.1" W
    V = 70%; Q = 20%; PZ = 0%

Stop #19-19 (elev = 4199'): 44° 30' 33" N, 108° 15' 23.9" W
    V = 95%; Q = 1%; PZ = 2-3%

Stop #21-19 (elev = 4157'): 44° 30' 33" N, 108° 15' 51.2" W
    V = 40%; Q = 50%; C = <1%; PZ = <1%

Stop #22-19 (elev = 4468'): 44° 31' 02.18" N, 108° 25' 13.16" W
   V = 80%; Q = 10%; PZ = 3%

Stop #24-19 (elev = 4672'): 44° 27' 43.7" N, 108° 31' 52" W
   V = 40%; Q = 50%; S = 5%; PZ = 3-5%

Stop #25-19 (elev = 4812'): 44° 27' 53.8" N, 108° 38' 21.5" W
   V = 40%; Q = 50%; S = 1%; PZ = 2-3%

Stop #3-21 (elev = 4699'): 44° 27' 51.9" N, 108° 34' 13.27" W
   V = 80%; Q = 8%; C = 5%; PZ = 6%

Stop #4-21 (elev = 4735'): 44° 27' 45.78" N, 108° 34' 19.04" W
   V = 80%; Q = 15%; C = trace; PZ = 2%

Stop #5-21 (elev = 4711'): 44° 28' 12.82" N, 108° 30' 38.98" W
   V = 85%; Q = 12%; C = 1%; PZ = trace

Stop #21-21 (elev = 4287'): 44° 30' 40.23" N, 108° 13' 38.02" W
   V = 80%; Q = 10%; C = 6%; PZ = 0%

Stop #22-21 (elev = 4210'): 44° 30' 32.9" N, 108° 13' 30.6" W
   V = 90%; Q = 5%; C = 1%; PZ = 0%

Stop #23-21 (elev = 4184'): 44° 30' 29.72" N, 108° 13' 29.21" W
   V = 85%; Q = 6%; C = 1%; PZ = 0%

Stop #7-22 (elev = 4765'): 44° 28' 47.6" N, 108° 32' 47.5" W
   V = 80%; Q = 15%; PZ = 3% (two cobble-size dolomites)

OTHER SURFACES

Stop #2-22 Bridger Butte SW (elev = 4977'): 44° 29' 23.6" N, 108° 32' 28.7" W
   V = 85-90%; Q = 12-18%

Stop #3-22 Bridger Butte NE (elev = 4965'): 44° 29' 36.0" N, 108° 32' 11.2" W
   V = 90%; Q = 10%

Stop #4-22 Bridger Butte north (elev = 4968'): 44° 29' 30.5" N, 108° 32' 23.9" W
   V = 95%; Q = 5%

Stop #6-22 lower terrace just west of Bridger Butte (elev = 4809'): 44° 29' 21.3" N, 108° 32' 51.4" W
   V = 88-90%; Q = 10-12%

Stop #8-22 High stripped surface between McCullough Peaks pediment and Bridger Butte (elev = 4975'): 44° 29' 29.2" N, 108° 37' 34.1" W
   No stream terrace deposits; some loess

TABLE V

Calculated lattice parameters and strains for quartz collected from Shoshone/Sunlight/Abiathar Detachment Fault breccia at the Dead Indian Summit Overlook (see Appendix I).

Sample	a (Å)	c (Å)	Strain ‖ a	Strain ‖ c
SiO_2	4.9014 (20)*	5.3918 (20)	-0.0024 (4)	-0.0025 (4)
$CaCO_3$	4.8741 (6)	5.3808 (6)	-0.0080 (1)	-0.0045 (1)
Control	4.9178 (8)	5.4075 (8)	0.0009 (2)	0.0004 (1)

*Numbers in parentheses are the standard 1σ errors in the final digits of the value

APPENDIX

Diffraction data and results from two quartz-bearing samples from fault gouge at the Chief Joseph turnout (Wyoming Hwy 296 = Chief Joseph Scenic Byway) on the Dead Indian Summit Overlook, at 44° 44' 37' N, 109° 22' 58" W, Park County, WY.

Methods

Diffraction data was collected for two quartz-bearing samples within the fault gouge as well as a control sample from a fossil-bearing micrite away from the fault. For each sample, 7 diffraction peaks corresponding to α-quartz were chosen for strain analysis (Miller indices [100], [011], [110], [102], [111], [200], [201]). Each peak was modeled with the Jade 2010 software package using a pseudo-Voight function with linear background, and peak parameters were refined using a least squares regression to find the best fit *d*-spacing. Next, the 7 resulting *d*-spacings were used to do a least squares regression of the quartz lattice parameters (*a, c*) for that sample. The best fit lattice parameters were then compared to the zero-strain lattice parameters at room pressure and temperature given by LePage and Donnay (1976) (a=4.9134 Å, c=5.4052 Å). The strain in each crystallographic direction was computed as (*a-a0*)/*c0*, respectively.

Results

Both samples from within the fault gouge showed significant negative (compressional) strain consistent with shearing during fault movement, whereas the control sample showed only a slight positive deviation from the zero-strain lattice parameters. The small degree of expansion observed away from the fault can be easily explained by the presence of significant concentrations of typical quartz impurities such as Al^{3+} and Fe^{3+} (Götze, 2009), whereas the strong compressional strain is only attributable to the shear stress along the fault zone. Table V shows the final results of the XRD analysis, including the relative strain.

Cover Image: View to east of Heart Mountain at sunset from Monument Hill Road.

Figure Captions

Figure 1. Google Earth image of Bighorn Basin, northwest Wyoming, showing several physiographic features discussed in text. Red dot east of the Y-U Bench is the approximate point of capture of Dry Creek by the Greybull River. See also **Figure 50**. Image modified from that of Lundeen and Kirk (2023).

Figure 2. A- View to west of relationships of allochthonous rocks of Shoshone/Sunlight/Abiathar Detachment Fault on a section of Cathedral Cliffs. Twp = lower and middle Eocene Wapiti Formation; Tcc = lower Eocene Cathedral Cliffs Formation; Mm = Mississippian Madison Formation; MDtj = Mississippian and Devonian Jefferson and Three Forks formations; Ob = Ordovician Bighorn Dolomite; STD = Surface of tectonic denudation. Reef Creek fault block is composed of allochthonous Mississippian Madison Formation. **B-** SSADF at Cathedral Cliffs, showing detail of relationship of surface of tectonic denudation (STD) to upper and lower plates. Symbols same as for Figure 2A. Cgc = Cambrian Grove Creek Formation; Csr = Cambrian Snowy River Formation. **C-** View to southwest of surface of tectonic denudation covered by volcanics and volcaniclastics of lower and middle Eocene Wapiti Formation (Twp) in area of Jim Smith Peak; image taken from 44° 55.5912' N, 109° 43.2902' W. Cp = Cambrian Pilgrim Formation; Ob = Ordovician Bighorn Dolomite.

Figure 3. A- View approximately to south of SSADF allochthon on "ramp" of Pierce (1957, 1960). Cp = autochthonous Cambrian Pilgrim Formation; Ob = Ordovician Bighorn Dolomite; MDtj = Mississippian and Devonian Jefferson and Three Forks formations; Mm = Mississippian Madison Formation; STD = Surface of tectonic denudation. **B-** View to west across Sunlight Basin showing light-colored allochthonous masses of Paleozoic carbonate rocks on surface of tectonic denudation (White Mountain, Steamboat, and rocks south of Steamboat), and on ramp (cliff in foreground, white arrow), as seen from rise above Chief Joseph Turnout off Wyoming State Highway 296. Twp = lower and middle Eocene Wapiti Formation.

Figure 4. View to west-northwest of HMMPS breccia of Bighorn Dolomite at massive block southeast of East Peak, McCullough Peaks. Hammer is 40.64 cm (16 in) long. Image by Mark Mathison.

Figure 5. View approximately to south of Pierce's (1960, 1980) "breakaway" area of Shoshone/Sunlight/Abiathar Detachment Fault. Tlcc = interbedded lower Eocene Lamar River and Cathedral Cliffs Formations; Twp = lower and middle Eocene Wapiti Formation. Solid line = Breakaway fault. Arrow shows direction of breakaway of Paleozoic rocks (out of image).

Figure 6. View to west of block of Devonian rocks of Jefferson and Three Forks formations (Djt) juxtaposed against block of Mississippian Madison Formation (Mm) forming allochthonous masses of Shoshone/Sunlight/Abiathar Detachment Fault lying on former (lower Eocene) land surface at approximately 44° 42' 19" N, 109° 21' 01" W. Site is interpreted to be part of source area for Heart Mountain/McCullough Peaks Sturzstrom deposit.

Figure 7. High-angle oblique Google Earth view to north-northwest of lower and middle Eocene volcanic and volcaniclastic rocks of Wapiti Formation (Twp) and probable equivalents of Aycross Formation (?Ta) exposed on Jim Mountain in Wapiti Valley, north side of North Fork of Shoshone River, west of Cody, WY. Note complete absence of any large normal or high-angle listric faults predicted by Hauge's continuous allochthon model (See also **Figure 8**).

Figure 8. View to north-northwest of lower and middle Eocene volcanic and volcaniclastic rocks of Wapiti Formation (Twp) lying atop non-volcanic lower Eocene Willwood Formation (Tw) on Jim Mountain in Wapiti Valley, north side of North Fork of Shoshone River, west of Cody, WY. Note complete absence of any large normal or high-angle listric faults predicted by Hauge's continuous allochthon model (See also **Figure 7**).

Figure 9. Interpretive sketches made by T.M. Bown in 1981 of chaotic deformation of rocks of middle Eocene Tepee Trail and Wiggins formations (Tttw) at **(A)** the "Rhodes allochthon" of Bown and Love (1987) and **(B)** in the upper valley of Twentyone Creek. Both show displaced masses of the Enos Creek/Owl Creek Debris-avalanche (ECOCDA) deposit, Hot Springs County, WY. Twl = lower Eocene Willwood Formation; Tac = lower and middle Eocene Aycross Formation; T-T' = thrust fault; D-D' = contact of ECOCDA displaced mass with *in situ* rocks. Note similarity of deformation to that of Decker's (1990) "chaotic dismemberment" of volcanic and volcaniclastic strata in upper valley of Shoshone River (our **Figures 10** and **11**) and near Rose Butte (**Figure 12**).

Figure 10. View to north of chaotically dismembered volcanic and volcaniclastic rocks, possibly of Deer Creek Member of Wapiti Formation (Twpd?), exposed on north side of South Fork of Shoshone River. Twp = Wapiti Formation. Note absence of any large normal/listric faults predicted by Hauge's continuous allochthon model (See also **Figures 11 and 12**).

Figure 11. View to north of chaotically dismembered volcanic and volcaniclastic rocks, probably of Deer Creek Member of Wapiti Formation, Wapiti Valley, north of North Fork of Shoshone River. Compare internal deformation with that in parts of Enos Creek/Owl Creek Debris-avalanche (**Figure 9**).

Figure 12. Chaotically deformed volcanic and volcaniclastic rocks probably equivalent to lower and middle Eocene Aycross (Ta?) Formation or Deer Creek Member of Wapiti Formation exposed southwest of Rose Butte, upper valley of Greybull River above Pitchfork Ranch, at approximately 44° 10' 47" N, 109° 14' 43'W. Google Earth image.

Figure 13. View to west-northwest of imbricated, differently inclined masses of allochthonous Paleozoic carbonate rocks at southeast edge of Sheep Mountain, north side of South Fork of Shoshone River, west of Cody, WY. Ob = Ordovician Bighorn Dolomite; Mm = Mississippian Madison Limestone.

Figure 14. **A-** View to north of imbricated, differently inclined masses of Bighorn Dolomite (Ob) on south side of White Mountain. **B-** Field sketch of wider view of imbricated Ordovician strata at same site.

Figure 15. Shingled, "piggy-backed" masses of variably dipping Mississippian Madison Limestone and older Paleozoic carbonate rocks developed on southeast side of Sheep Mountain, north side of South Fork of Shoshone River, west of Cody, WY. Such relationships do not accord with Hauge's hypothesis of gravitational spreading of allochthonous Paleozoic blocks following deposition of Wapiti volcanic and volcaniclastic rocks. Twp = lower and middle Eocene Wapiti Formation; Qls = Quaternary landslide deposits.

Figure 16. View approximately to north of contact relationships of allochthon and autochthon of Shoshone/Sunlight/Abiathar Detachment Fault (SSADF) at White Mountain, Sunlight Basin, Absaroka Range, northwest WY. DS = Detachment fault surface (surface of tectonic denudation); Csr = Cambrian Snowy River Formation; Oba = Ordovician Bighorn Dolomite on autochthon; Obuc = Ultracataclastite gouge at base of allochthon; Obm = Partially marbleized Ordovician Bighorn Dolomite forming lower part of SSADF allochthon.

Figure 17. View approximately to east-northeast of exposures at west end of White Mountain, Sunlight Basin, showing: allochthonous SSADF rocks of partially marbleized Bighorn Dolomite (Ob); diorite stock, lamprophyre dike, and trachyandesite; and overlying lower and middle Eocene volcanic and volcaniclastic rocks of Wapiti Formation (Twp). Malone *et al.* (2017) suggested that the lamprophyre intrusion, of which this dike is an allochthonous remnant, was instrumental in triggering the HMDF.

Figure 18. Close-up image of outcrop of transported SSADF upper plate remnant of lower Eocene Crandall Conglomerate at approximately 44° 43' 17" N, 109° 31' 41" W, in valley of Beem Gulch, Absaroka Range. Drone image by Mark Mathison.

Figure 19. **A-** View to east of lower Eocene Crandall Conglomerate (Tec) lying in channel cut into Cambrian rocks overlying Pilgrim Formation (Cp) above Squaw Creek, Absaroka Range. Both units are part of autochthon of Shoshone/Sunlight/Abiathar Detachment Fault. **B-** East face of channel deposit of lower Eocene Crandall Conglomerate (Tec) depicted in **Figure 19A**. Drone image by Mark Mathison.

Figure 20. View to south-southwest of massive Paleozoic carbonate sturzstrom allochthons on Heart Mountain, north-northwest of Cody, WY. Ob = Bighorn Dolomite (Ordovician); MDtj = undifferentiated Jefferson and Three Forks formations (Mississippian and Devonian); Mm = Madison Formation (Mississippian); cld = Comminuted landslide debris. All units shown are allochthonous and form part of the Heart Mountain/McCullough Peaks Sturzstrom deposit.

Figure 21. View to east-northeast of west side of Heart Mountain, showing displaced rocks of Heart Mountain/McCullough Peaks Sturzstrom (HMMPS). Ob = Ordovician Bighorn Dolomite; MDtj = Mississippian and Devonian Jefferson and Three Forks formations; Mm = Mississippian Madison Formation. Kme = *in situ* Upper Cretaceous Meeteetse Formation.

Figure 22. High-angle oblique Google Earth view to northwest of dissected McCullough Peaks pediment (TQp) developed on lower Eocene Willwood Formation (Tw). Red arrows highlight exposed edges of veneer of light-colored HMMPS carbonate debris beneath darker vegetated pediment surface.

Figure 23. High-angle oblique Google Earth image approximately to south-southeast of allochthonous masses of brecciated Paleozoic carbonate rock (HMMPS) in the vicinity of East Peak of McCullough Peaks. Ob = Massive triangular block of Ordovician Bighorn Dolomite discussed in text.

Figure 24. View to south of fragmented mass of Ordovician Bighorn Dolomite littering surface of McCullough Peaks pediment (TQp). Mass appearance is typical of larger preserved remnants of HMMP sturzstrom deposit.

Figure 25. View approximately to east of Corbett Mass #2 (left and middle HMMPS designations on image; see Table II), atop truncated lower Eocene Willwood Formation (Tw) above valley of Shoshone River, from U.S. Highway 14A (See **Figures 26-28**).

Figure 26. View to northeast of Corbett Mass #2 (Table II), showing mass of disaggregated allochthonous debris and welded breccias of Paleozoic carbonate rocks (mixed Ordovician Bighorn Dolomite and Mississippian Madison micrites) of HMMPS lying atop erosional surface on lower Eocene Willwood Formation (Tw). Shoshone River valley on skyline in upper left. Drone view from Mark Mathison.

Figure 27. View approximately to east of remnants of Heart Mountain/McCullough Peaks Sturzstrom (HMMPS) deposit forming Corbett Mass #2 in foreground (see Table II), lying atop eroded surface on lower Eocene Willwood Formation (Tw). Drone view from Mark Mathison.

Figure 28. View approximately to northwest of welded breccia of Paleozoic carbonate clasts as exposed on Corbett #2 HMMPS deposit. Hammer is 40.64 cm (16 in) long. Image by Mark Mathison.

Figure 29. High-angle oblique Google Earth image to east-southeast across the McCullough Peaks badlands, showing West Peak (Pyramid Mountain), Middle Peak (McCullough Peaks HP on image), and East Peak of McCullough Peaks eminence. Tw = lower Eocene Willwood Formation. Most of the preserved part of the McCullough Peaks pediment (Figure 22) lies out of the image to the east.

Figure 30. View to east of weathered roadcut exposure of SSADF or HMMPS breccia exposed on east side of Wyoming State Highway 296, opposite Chief Joseph turnout. Mark Mathison (left) and Tom Bown (right) for scale.

Figure 31. Specimen of breccia of gouged Ordovician Bighorn Dolomite taken from roadcut outcrop, east opposite Chief Joseph on Wyoming State Highway 296, east of descent into Sunlight Basin, Absaroka Range (See **Figure 30** and Table V). It is unclear if this breccia was created by movement over the ramp by an allochthon of the SSADF, or by initial landsliding movement on the HMMPS.

Figure 32. Close-up view to northwest of brecciated Ordovician Bighorn Dolomite (Ob) forming part of ~17,150 m^3 allochthonous mass at 44° 34' 21" N, 108° 49' 04" W, about 750 m southeast of East Peak of McCullough Peaks. Tw = lower Eocene Willwood Formation. See also Figure 23 for location.

Figure 33. View approximately to northeast of boulders, cobbles, and other debris of Paleozoic carbonate rocks making up probable allochthonous HMMPS debris in canyon of Rattlesnake Creek (see Figures 1, 44, and 50 for location), north of Buffalo Bill Reservoir in Wapiti Valley, west of Cody, WY. Geologist (arrow) for scale.

Figure 34. Geologist Al Warner on surface near most distal pediment remnant covered with unsorted granule-to-boulder dolomitic debris of Heart Mountain/McCullough Peaks Sturzstrom derived from Ordovician Bighorn Dolomite. View approximately to southwest from 44° 45' 16" N, 109° 03' 26" W, Park County, WY. Photo by T.M. Bown.

Figure 35. Angular Paleozoic carbonate debris and boulder forming part of easternmost preserved remnant of runout of Heart Mountain/McCullough Peaks Sturzstrom. Hammer is 40.64 cm (16 in) long. Image by Mark Mathison.

Figure 36. A- View to west-northwest, up the McCullough Peaks pediment surface, showing thickness of Paleozoic carbonate debris (about 6 m) exposed at margins of dissected pediment and patches of relatively darker green vegetation at seeps along contact of HMMPS debris and impermeable mudstones of the lower Eocene Willwood Formation (concealed beneath vegetation). **B-** View approximately to southeast showing inclined, dissected McCullough Peaks pediment surface (TQp) east of East Peak. HMMPS = Vegetated pediment surface with veneer of comminuted Paleozoic carbonate rock landslide debris.

Figure 37. View approximately to southwest of Heart Mountain, showing remnants of HMMPS carbonate debris-covered pediment surfaces (TQp) descending to the northeast.

Figure 38. View to north up the Heart Mountain pediment surface, showing projected confluence of pediment remnants (TQp) with pediment surface directly underlying massive blocks of bedded Paleozoic carbonate rocks of Heart Mountain/McCullough Peaks Sturzstrom (at tree line, marked by arrows). Image taken from 44° 36' 09.3" N, 109° 06' 10.6" W.

Figure 39. View to north of allochthonous Paleozoic rocks of HMMPS lying atop Tertiary/Quaternary pediment surface (TQp). Note that pediment surface in foreground is covered with comminuted landslide debris and rises to the north to become confluent with buried pediment at tree line marked with arrows (See also **Figure 38**).

Figure 40. View approximately to east of Heart Mountain pediment surface (TQp) on south side of Heart Mountain sloping upward to north to meet base of allocthonous HMMPS deposit (arrow) on Heart Mountain. See also **Figures 38 and 39**.

Figure 41. High-angle oblique Google Earth view to northwest of two late Quaternary or Holocene earthflows (Qef) on east side of Heart Mountain, on either side of 44° 41' 48" N, 109° 05' 10" W.

Figure 42. High-angle oblique Google Earth view of earthflow on north side of McCullough Peaks, at 44° 35' 26" N, 108° 49' 30" W. Tw = Lower Eocene Willwood Formation.

Figure 43. View approximately to northwest from HMMPS breccia outcrop at Corbett Mass #2, across valley of Shoshone River to Heart Mountain, Pat O'Hara Mountain, and Natural Corral. The Natural

Corral area was almost certainly the source of the Heart Mountain/McCullough Peaks Sturzstrom deposit—the largest landslide in the world. Image by Mark Mathison.

Figure 44. **A-** Sketch map of hypothetical reconstruction of area originally covered by debris of the Heart Mountain/McCullough Peaks Sturzstrom, in relation to regional physiographic features and present day towns. A = Easternmost preserved Paleozoic carbonate debris emplaced by HMMPS (see **Figures 35, 45, and 46**); B = northernmost preserved HMMPS Paleozoic carbonate debris on west side of Shoshone River; C = Paleozoic carbonate landslide debris that we tentatively attribute to the HMMPS in Rattlesnake Canyon (see **Figure 33**); D = Remnant of HMMPS deposit northeast of Carter Mountain (See below, and **Figure 56**); E = Northernmost preserved Paleozoic carbonate debris on east side of Shoshone River, probably emplaced by HMMPS. **B-** Comparison of run-out and areal cover of HMMPS with those of the Baga Bogd, Saidmarreh, and other landslides, and that of the Enos Creek/Owl Creek Debris-avalanche (ECOCDA; Enos Creek – Owl Creek in figure). Arrows show directions of displacements. Note that although the HMMPS and ECOCDA have similar run-out distances, the volume of and area covered by ECOCDA debris is far greater.

Figure 45. View approximately to southwest of high pediment remnant (Qp) capped by easternmost known preserved deposit of HMMPS (See also **Figures 35 and 46**).

Figure 46. View approximately to northeast of of high pediment remnant (TQp) capped by easternmost known preserved deposit of HMMPS (See also **Figures 35 and 45**). PB = Polecat Bench (in far distance). Twl = lower Eocene Willwood Formation.

Figure 47. Stripped surfaces in source area (background), and huge masses of displaced rock (foreground) of immense Flims, Switzerland, sturzstrom event.

Figure 48. View to southeast of dissected edge of McCullough Peaks pediment (TQp), showing allochthonous Paleozoic carbonate debris (HMMPS) littering surface, and spring seep (foreground) marked by dark green vegetation at contact of HMMPS debris with underlying impermeable Willwood mudstones.

Figure 49. View to south of Ruler Bench—remnant of a high-level pediment surface off the Beartooth Mountain massif south of Red Lodge, MT. We surmise that it was across a similar high-level, mountain-proximal surface that the HMMPS debris passed from its source to McCullough Peaks in the Bighorn Basin.

Figure 50. Sketch map of a part of the northwestern Bighorn Basin, showing relationships of several physiographic features discussed in text. PC = Point of capture of Dry Creek by the Greybull River = Point "D" of Mackin (1936). Note locations of CM (northeast of Carter Mountain) allochthon, Elk Butte, Oregon Basin drainage of Dry Creek, Meeteetse Rim, and Rattlesnake Canyon (discussed in text).

Figure 51. View to east across part of southern margin of Oregon Basin Dome, showing second highest (second oldest, after Elk Butte) terrace surface (Qt) in the area. Kco and Kmv = Upper Cretaceous Cody and Mesaverde formations. The terrace deposits contain clasts of Paleozoic carbonates derived from debris of the HMMPS.

Figure 52. View to west-southwest of Elk Butte in the Oregon Basin (see Figure 50), the top of which is capped by the highest terrace deposit (at 1,893 m) in the drainage of Dry Creek (Table IV). The terrace is rich in pebbles and cobbles derived from Paleozoic carbonate rocks. Kco = Upper Cretaceous Cody Shale.

Figure 53. View to east of West (proximal) and East (distal) of Squaw Buttes, the most distally preserved allochthonous masses of the Enos Creek/Owl Creek Debris-avalanche (Bown, 1982a,b; Bown and Love, 1987). Tt = non-volcanic sandstones, mudstones, shales, and oil shales of lower Eocene Tatman Formation; Tva = Allochthonous volcanic and volcaniclastic rocks of Aycross and younger strata displaced by Enos Creek/Owl Creek Debris-avalanche.

Figure 54. View to north of displaced, steeply dipping rocks of Aycross Formation entrained by movement on the Enos Creek/Owl Creek Debris-avalanche; upper drainage of Twentyone Creek, Hot Springs County, WY.

Figure 55. Map showing areal and topographic distribution of allochthonous masses of the Enos Creek/Owl Creek Debris-avalanche in the southeast Absaroka Range, showing relationship of masses to displaced volcanic and volcaniclastic debris atop Squaw Buttes (red dot) in the southwest Bighorn Basin.

Figure 56. View to east of allochthonous mass of disaggregated Paleozoic carbonate rocks (largely Bighorn Dolomite) at 44° 21' 19" N, 109° 10' 46" W, northeast of Carter Mountain (CM allochthon on map, **Figure 50**).

Figure 57. High-angle oblique Google Earth image of view approximately to north of area at juncture of Rattlesnake Mountain and Pat O'Hara Mountain. We interpret this area as having been the source of HMMPS Paleozoic carbonate debris northeast of Carter Mountain (See Figures 50 and 55). PZ allochthon = SSADF allochthon of steeply dipping Paleozoic carbonate rock; Trc = Triassic Chugwater Formation; Twl = lower Eocene Willwood Formation; Tac = lower and middle Eocene Aycross Formation.

Figure 58. View to north of SSADF allochthon of steeply dipping Paleozoic carbonate rocks in area of juncture of Rattlesnake Mountain and Pat O'Hara Mountain, Park County, WY (See Figure 56). PZ = SSADF allochthon of Paleozoic carbonate rocks; Twl = lower Eocene Willwood Formation.

Figure 59. Bill Pierce (left) and Tom Bown (right) leaving field in Twentyone Creek area of Enos Creek/Owl Creek Debris-avalanche in 1979. Photograph by Dave Love.

Figure 60. Dave Love on Carter Mountain, 1979; photograph by T.M. Bown.

Figure 61. Dr. Carl F. Vondra; photograph by Mark E. Mathison.

Figure 62. Mark Mathison in ancient Deir Abu Lifa monastery, Fayum Depression, Egypt.

Figure 63. Tom Bown (left) and Al Warner (right) examining maps at DDX Ranch, 2021.

Figure 1

Figure 2A

Figure 2B

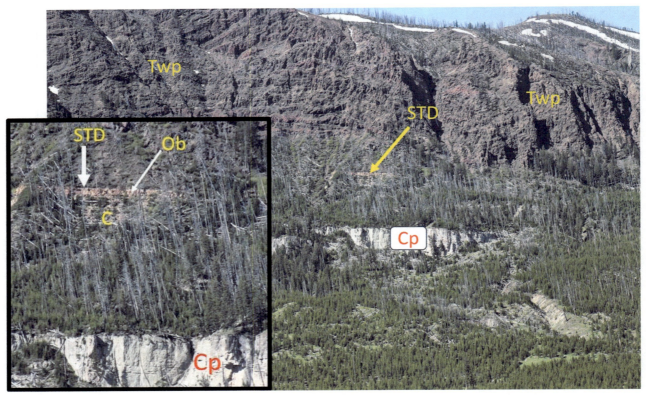

Figure 2C

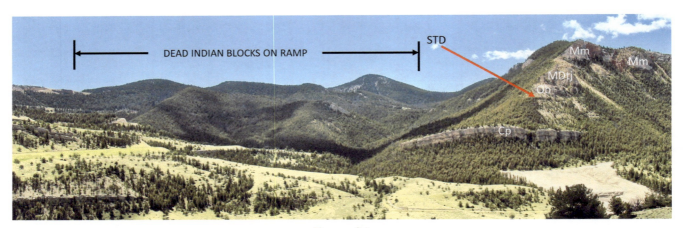

Figure 3A

Figure 3B

Figure 4

Figure 5

Figure 6

Figure 7

Figure 8

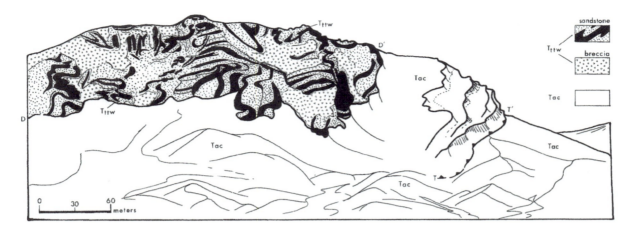
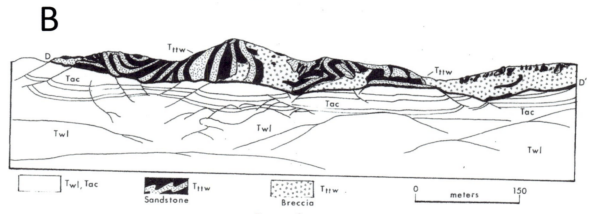

Figure 9

Figure 10

Figure 11

Figure 12

Figure 13

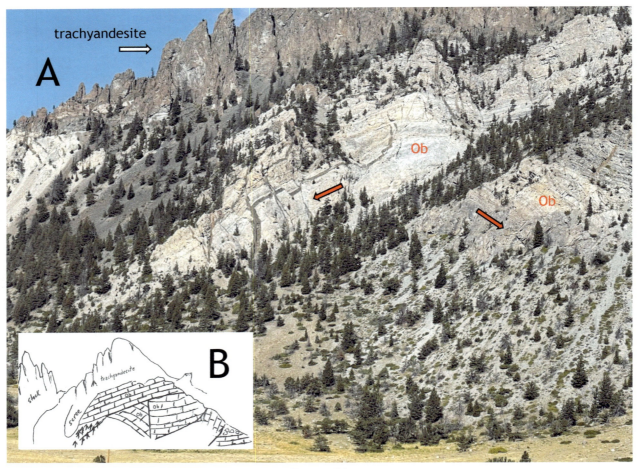

Figure 14

Figure 15

Figure 16

Figure 17

Figure 18

94

Figure 19A

Figure 19B

Figure 20

Figure 21

Figure 22

Figure 23

Figure 24

Figure 25

Figure 26

Figure 27

99

Figure 28

Figure 29

Figure 30

Figure 31

Figure 32

Figure 33

Figure 34

Figure 35

Figure 36A

Figure 36B

Figure 37

Figure 38

Figure 39

Figure 40

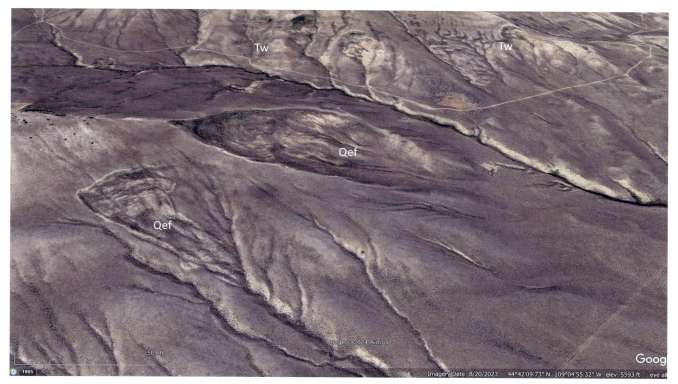

Figure 41

Figure 42

Figure 43

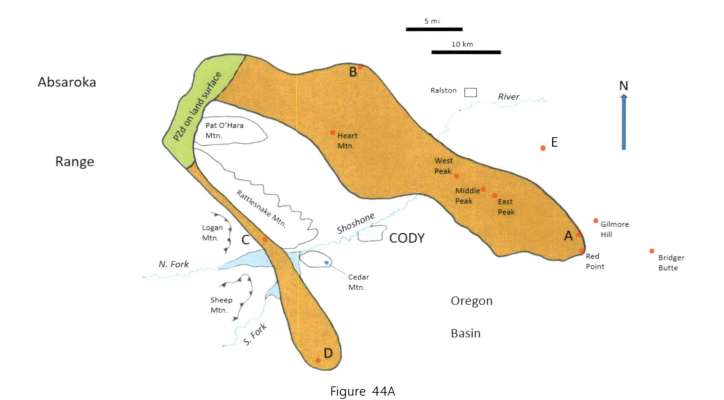

Figure 44A

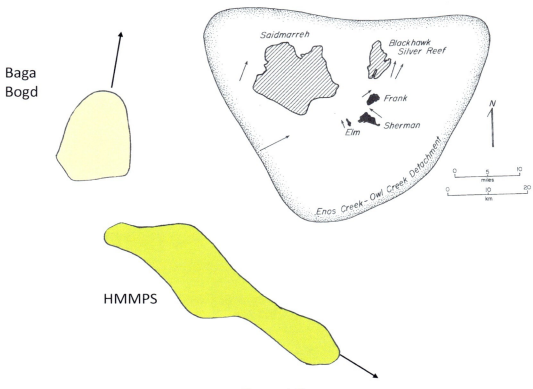

Figure 44B

Figure 45

Figure 46

Figure 47

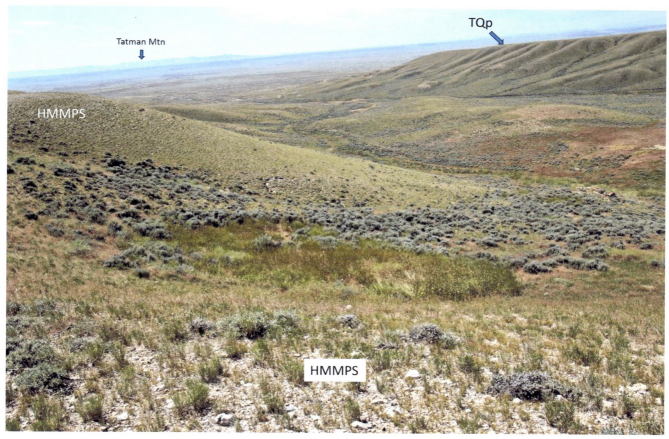

Figure 48

Figure 49

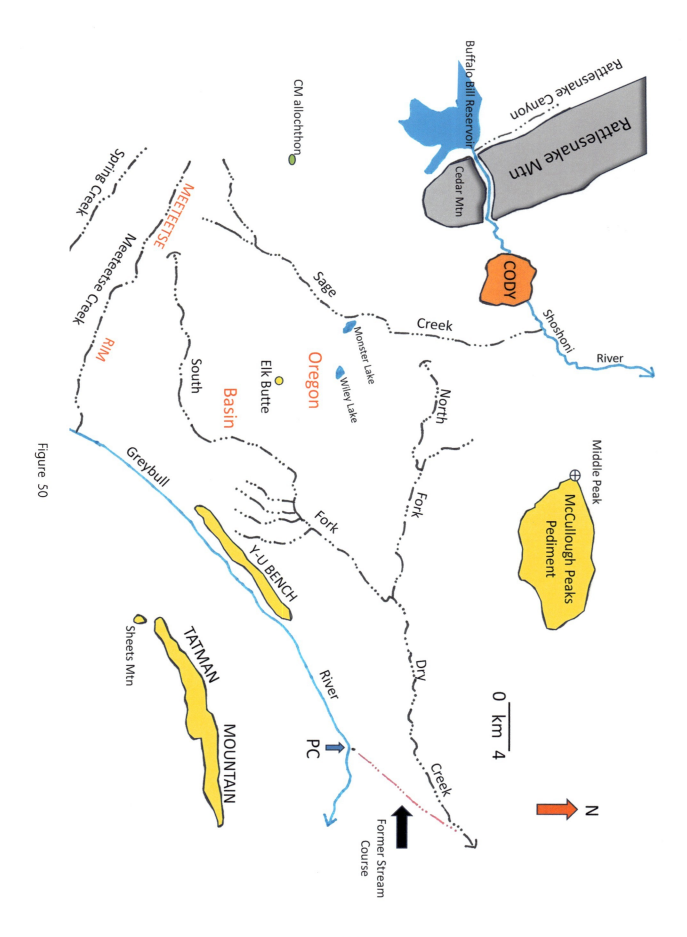
Figure 50

Figure 51

Figure 52

Figure 53

Figure 54

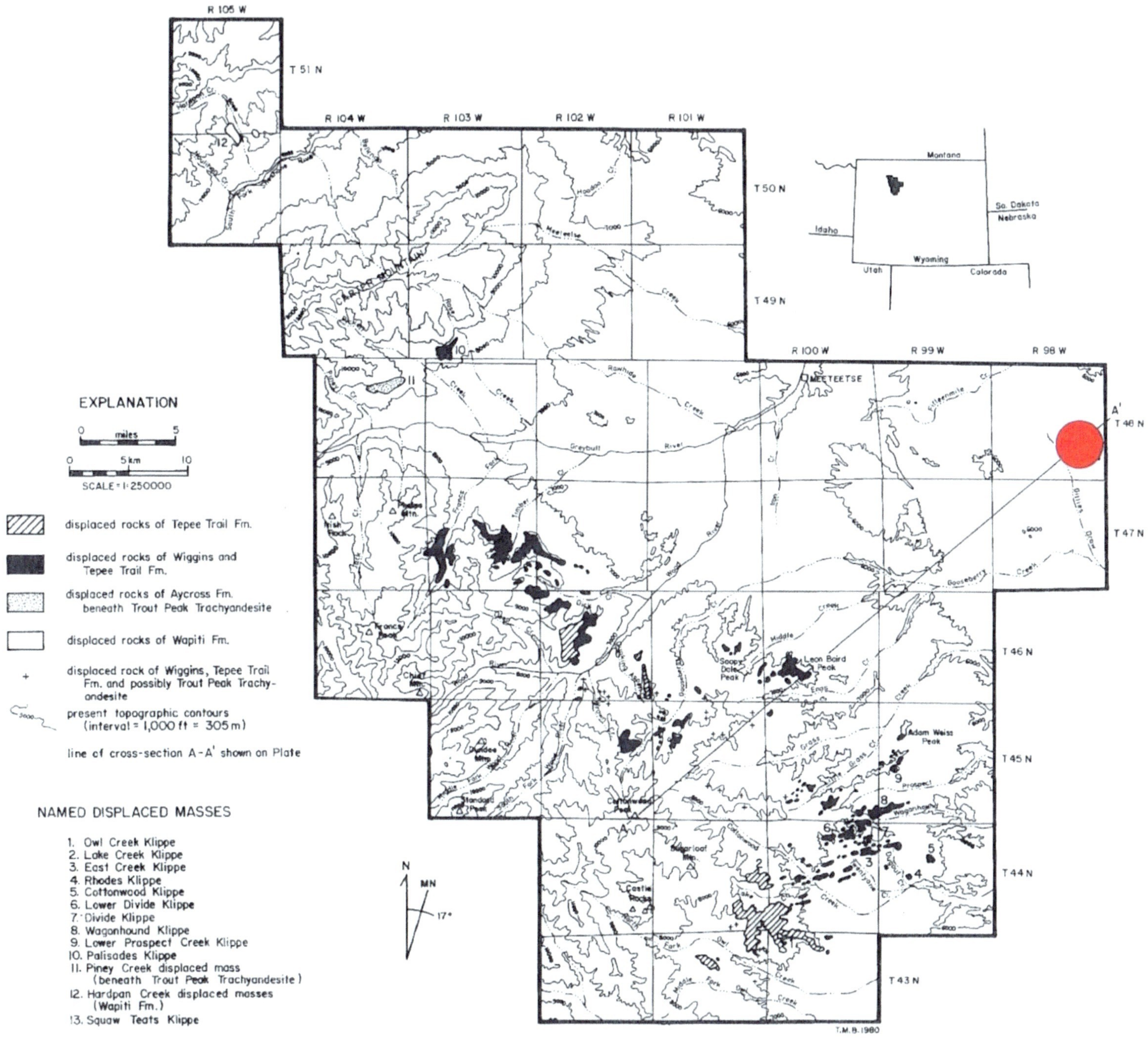

R 105 W
T 51 N
R 104 W
R 103 W
R 102 W
R 101 W
T 50 N
T 49 N
R 100 W
R 99 W
R 98 W
T 48 N
T 47 N
T 46 N
T 45 N
T 44 N
T 43 N
A'
A
MEETEETSE
Greybull
River
Rawhide
Creek
Meeteetse
Hoodoo Cr.
Fifteenmile
Gooseberry
Creek
Leon Baird Peak
Soapy Dale Peak
Adam Weiss Peak
Sugarloaf Mtn.
Castle Rocks
Francs Peak
Chief Mtn.
Irish Rock
Dundee Mtn.
Standard Peak
Cottonwood Peak
Montana
Idaho
Wyoming
Utah
Colorado
So. Dakota
Nebraska
N
MN
17°
T.M.B. 1980
EXPLANATION
0 miles 5
0 5 km 10
SCALE = 1:250000
displaced rocks of Tepee Trail Fm.
displaced rocks of Wiggins and Tepee Trail Fm.
displaced rocks of Aycross Fm. beneath Trout Peak Trachyandesite
displaced rocks of Wapiti Fm.
displaced rock of Wiggins, Tepee Trail Fm. and possibly Trout Peak Trachyandesite
present topographic contours (interval = 1,000 ft = 305 m)
line of cross-section A-A' shown on Plate
NAMED DISPLACED MASSES
1. Owl Creek Klippe
2. Lake Creek Klippe
3. East Creek Klippe
4. Rhodes Klippe
5. Cottonwood Klippe
6. Lower Divide Klippe
7. Divide Klippe
8. Wagonhound Klippe
9. Lower Prospect Creek Klippe
10. Palisades Klippe
11. Piney Creek displaced mass (beneath Trout Peak Trachyandesite)
12. Hardpan Creek displaced masses (Wapiti Fm.)
13. Squaw Teats Klippe

Figure 55

Figure 56

Figure 57

Figure 58

Figure 59

Figure 60

Figure 61

Figure 62

Figure 63

ABOUT THE AUTHORS

Albert Warner is a petroleum geologist who received his B. Sci. (1965) and M. Sci. (1968) in geology at Iowa State University, and his Ph. D. in geology at the University of Iowa (1978). His professional career began in 1968 with Shell Oil Company in the Permian and Palo Duro Basins of Texas and New Mexico followed by 3 years prospecting in the Appalachian Basin for Columbia Gas Transmission Corporation. After spending 4 years from 1973-1977 earning his Ph. D. he returned to the petroleum industry in Oklahoma City where he developed drilling prospects in the Pennsylvanian sandstones of the Anadarko Basin for a number of companies including Gulf Oil Corporation, Slawson Oil, and Chesapeake Energy as well as an independent consulting geologist ultimately finding nearly 300 million barrels of oil equivalent. Additionally, he began a basin wide study of the principal producing Pennsylvanian sandstone reservoirs in the Anadarko Basin that has continued for 4 decades. As a member of the AAPG he has twice presented papers at annual regional meetings. Al is preparing a photoglossary of the Heart Mountain complex, a subject that has interested him for more than 50 years. awarner43@gmail.com

Thomas Bown is a geologist/paleontologist who received his B. Sci. in geology at Iowa State University in 1968 and his Ph.D. in Geology at the University of Wyoming in 1977. He has accomplished geologic field studies in Wyoming's Bighorn Basin for 60 years, in Egypt (22 seasons), Argentina (12 seasons), the UAE (3 seasons), Ethiopia (3 seasons), and in 6 other foreign countries. While working for Yale's Peabody Museum in 1969, he and two others drove a field vehicle from India to Libya. At the U.S. Geological Survey and as a consulting geologist he contributed over 270 scientific publications and more than 100 technical site reports on topics as diverse as sedimentology, fossil mammals and trace fossils, paleosols, ground water resources, taphonomy, the world's oldest paved road, the world's oldest cattle kraals, and the Enos Creek/Owl Creek Debris-avalanche. He was the primary consultant on Britain's Channel 4 documentary on The Lost Army of King Cambyses, and he published an historical novel, *The Efreet*, on the latter topic. Awards include the Outstanding Paper Award (SEPM), The Meritorious Service Award (U.S. Department of the Interior), The Morris F. Skinner Award (Society of Vertebrate Paleontology), The Distinguished Alumnus Award (Iowa State University), and a cash award from the United Arab Emirates for Outstanding Contributions to their Groundwater Research Program. kanprimate@aol.com

Mark Mathison is a geologist/paleontologist/geomorphologist/stable isotope geochemist who received his B. Sci. in geology at Iowa State University in 1995 and his M.S. In geology also at Iowa State University in 2000. He has accomplished geologic field studies in Wyoming's Bighorn Basin for 25 years, in Egypt (12 years), Norway (3 seasons), Sweden (2 seasons), Ethiopia, Mongolia, and in 5 other foreign countries. His work in Ethiopia resulted in the discovery of a new species of fossil fox (Vulpes mathisoni) that he had the honor to have named after him. 24 of the 25 years of work in Wyoming's Bighorn Basin also included lecturing and managing Iowa State University's geology field station near Shell, WY. Hundreds of field geologists were produced during this time. He was part of two influential papers; one on glacier slip and seismicity induced by surface melt, and a second on basal anthropoid primates from Egypt and the antiquity of Africa's higher primate radiation. Mark is a commercial drone pilot and uses this skill to assist in geological research and to detect fossils by ultraviolet fluorescence by drone. With his technical skills he has developed wireless sensors to detect slip at the base of glaciers and has helped develop other automated sensors for use in geological fieldwork. mathison@iastate.edu